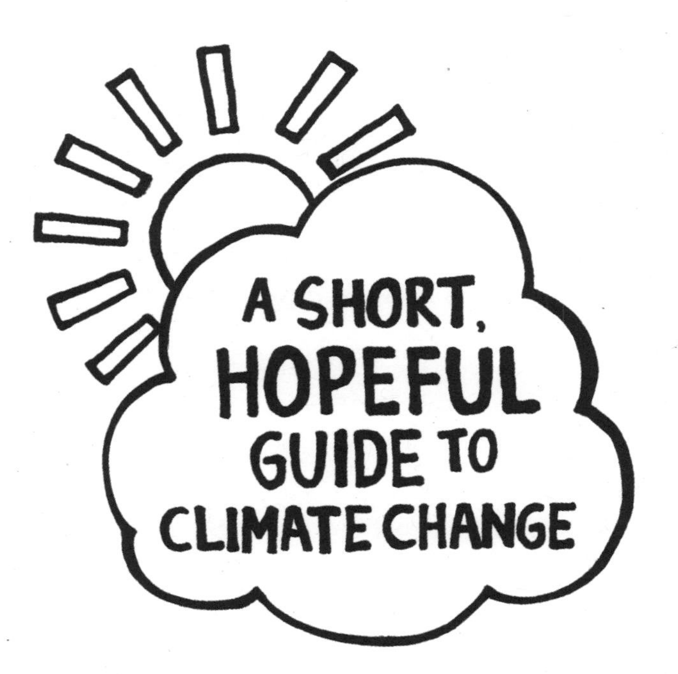
A SHORT,
HOPEFUL
GUIDE TO
CLIMATE CHANGE

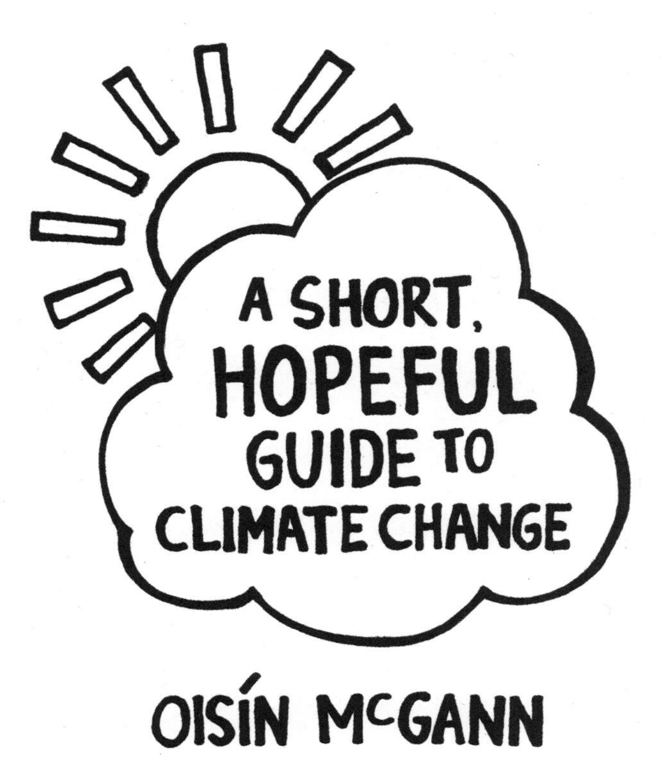

OISÍN McGANN

Published by Little Island Books
in association with Friends of the Earth Ireland

Eco-friendly printing by Ashley House Printing Company

ashley house
printing company

A SHORT, HOPEFUL GUIDE TO CLIMATE CHANGE

First published in 2021 by
Little Island Books
7 Kenilworth Park
Dublin 6W
Ireland

A British Library Cataloguing in Publication record for this book is available from the British Library.

Cover and interior illustrations by Oisín McGann
Designed by James Tuomey
Proofread by Emma Dunne
Printed in the UK by Ashley House Printing Company

Print ISBN: 978-1912417742
Ebook (Kindle) ISBN: 978-1912417803
Ebook (other platforms) ISBN: 978-1912417810

Little Island receives funding from the Arts Council of Ireland/ An Chomhairle Ealaíon

10 9 8 7 6 5 4 3 2 1

CONTENTS

FOREWORD

Climate change is a 'wicked' problem, mostly because our civilisation is based on fossil fuels. If we fail to solve this problem, the impacts of global warming will be like something out of an apocalyptic science fiction movie. But if, instead, we choose to take action, we could have a much healthier society than we have today, with a cleaner environment and better quality of life. We are now at the tipping point of this decision: whether to venture into a civilisation beyond fossil fuels or to continue with business as usual.

As a parent, I try to protect my daughter from scary stories, but the story of climate change is one I cannot protect her from. Whether we tip towards climate action or climate chaos, the world she will experience as an adult will be very different from mine. Either way, I need to prepare her, and this *Short, Hopeful Guide to Climate Change* is one way to do so. As her generation – your generation – becomes more aware of environmental issues and the limits of our planet's resources, you need clear information so you can make good decisions. And that is what you will find in this book.

We are fortunate to live in a time when all the solutions to climate change are available to us. It is simply a matter of choosing which ones to implement. Each solution has its pros and cons, and understanding those nuances is critical to deciding what kind of world we want for our future. That is why Oisín McGann presents you with a wide range of options and paints a picture of what the world could look like if we take sensible climate action. Each solution discussed here benefits both people and the planet, and compassion for the millions of other species we share this world with is a central theme of this guide.

This book blends its author's passion for science with his curiosity about human behaviour in a light-hearted take on a topic we usually think of as very heavy, so you will enjoy reading it as well as learning a lot about how interconnected we are with this pale blue dot in the universe that we call home. Anxiety about climate change is a normal response to the challenge we face, as Oisín explains it here, but this book shows you how you can channel your concern into collective action.

Young people are already charting the path forward in this global climate movement. The 'stories of change' presented in each chapter remind us that climate solutions are not just concepts for the future, but are actions that are happening all around the world today, demonstrating that humanity is already transitioning to a new civilisation. This short and hopeful guide helps us to realise that climate change may not be so 'wicked' after all, because we have real cause to hope that a better world is now within our grasp.

Dr Cara Augustenborg
Environmental scientist and broadcaster

INTRODUCTION
WHY THIS MATTERS TO YOU

We live in a world where humans can land a spacecraft on a comet – which is like trying to land a speeding bullet on another speeding bullet, only many times harder. Though I am not a scientist, I am a fan of science. To me, it is magic, but in real life. And what I am *most* interested in is that contact point where science meets human nature. Because humans are ... well, you know ... *weird.*

And, while this book is about climate change, it's about *our* human story too – all the acts of brilliance and absurdity that got us here, and where we're off to next. I want to share my sense of wonder at our existence. We're the species that invented cheese and nuclear weapons. We created the internet and then used it to post cat videos. We're living on a planet that is a freak oddity in our galaxy – and *you* are completely woven into it all.

Your environment is everything beyond your flesh. It immerses you in its protective atmosphere and presses its air against your skin. It fills your lungs. Your body is built out of the air you breathe and the food and drink

you consume. Planet Earth is your life-support system, and you are permanently plugged into it.

It may not surprise you to learn that there's some *really bad news* in this book. There's just no denying that humans have messed things up in a big way. And yet, within that mess lies hope for the future. Because a long time ago, we started outrunning evolution, racing ahead of the natural world, and it has led us to where we are now. Like a rocket leaving scorch marks and smoke behind it, we didn't pay enough attention to the damage we left in the wake of our progress. But we're paying attention *now* – now that we've become aware of our power. And, to quote Spiderman (and lots of other people before him), with great power comes great responsibility.

I'm not writing about climate change to inspire fear or guilt. No: I want to share my fascination with it. The changes that are happening around us are revealing how beautifully complex and interconnected our planet is. There is so much that is striking and strange and wild about it that, no matter what your interests and passions are, there is something that will light your fire.

Speaking of fire, we talk a lot about the negatives of fossil fuels and how they've affected our atmosphere, but we also have to appreciate what they have given us: the energy to create our advanced civilisation. We wouldn't be where we are today without wood, peat, coal, gas and oil. We can carry a million pages of text on a USB drive that fits on a keyring, transplant a new heart into a living human being or film a giant squid at the bottom of the ocean – all of which is pretty good, and all of which we can do because of the way we have used the world's resources so far.

And it's not just our science that has progressed over this time. We've developed as civilised beings too. As our world becomes more unified, more enlightened and better educated, we have established ideas like human rights, democracy and the value of empathy. We are taking better care of each other. Our world is more peaceful than it has ever been.

But we have burned our way to the top. Progress and pollution are twisted together in the story of the human race. Our development has come at a huge cost to the thin, living skin wrapped around this ball of rock we call Earth.

This is an issue that affects everyone, everywhere, but I live in Ireland, and I am conscious that there will be points in the book where I refer to things that 'we' have; in other words, I'm referring to a normality that I live in, but that not all other people around the globe will share. Our world is an extraordinarily varied place. Though I write from my own experience of life, it's only one point of view. There are countless others with very different lives and experience, and indeed, many of the countries that are being affected most by climate change have contributed the least to its advance. Industrial development has caused this problem, and the most developed nations, including mine, bear the greatest responsibility to do something about it.

The most important thing I want you to take away from this book is that we are *already* transforming our civilisation. Society is in a constant state of change, and the same progress that has left its scorch marks on our environment is steadily carrying us away from our fossil-fuelled past.

You cannot solve climate change all by yourself, and nobody is expecting you to. Billions of people created this crisis over centuries, through countless tiny actions, and so the sheer enormity of it can feel overwhelming. Throughout the book, I will describe the different changes taking place in our environment and, while it may feel as if the scale of these changes is beyond our power to overcome, you should always remember that *humans caused them*. And we did it by *accident*. Now, however, we are waking up to what we've done, and in the greatest movement ever seen in human civilisation, people around the world are grasping the *power* that caused this. Through a myriad of small actions, they are redirecting that power.

No single person or organisation or country can slow the changes in our climate, and yet billions of people taking tiny actions every day can do that and more, helping steer us towards becoming a wiser, healthier and more sustainable society. This is a big ship, but if we all contribute, if we all pull the same way, we can turn it around. It is an unprecedented situation, and one I find both thrilling and inspiring. The human race, in every country around the globe, and at every level of society, has begun to unite to take on a planet-sized problem.

Climate change is a threat to everything we know, but the story of climate change is also the story of our civilisation. It is an epic and fascinating tale about our success as a species, and it is ongoing. That story is carrying us into the future: we are gaining increasing foresight and exerting more control over our path to that distant horizon, and there are as many ways to get involved and help solve this as there are problems to solve.

So as you read this book, always remember that the same mind-boggling power that changed the climate on our planet by accident can also be used *deliberately*, with compassion, intelligence and wisdom. That is our goal. It is an incredibly ambitious undertaking, but I believe that, in making a better world, we will make ourselves better human beings.

I have great hope for the future we are creating for ourselves.

CHAPTER I
PURE LUCK

It has to be said that what humans have done is pretty impressive. Like some super-villain in a Hollywood movie, we've managed to change our world's weather, except we did it entirely *without meaning to*. For too long, we've been getting ahead of ourselves, racing into the future, without looking at where it's taking us. In fact, it would be true to say that the story of climate change is also the story of human development. And that, as I will show over the course of this book, is a reason for hope.

Amazing though it might seem, we're not the first life forms on Earth to cause problems by changing the climate – but the first ones to do it literally had no brains, as you'll see in a minute. Fortunately, we *do* have brains, and we have some seriously intelligent and powerful people working to *fix* things.

I'll get to that later. First, I want to talk about how lucky we are that we exist at all, because this big blue ball we live on has done a lot of changing over the last four and a half billion years. That life-giving air that surrounds you, pressing against your skin and filling your lungs, and all that comes with it, took a while to show up on Earth. For a lot of those four and a half billion years, we wouldn't have been able to survive on this planet. So take a look around you, breathe in that sweet, sweet oxygen and be grateful that you're alive. Go ahead, I'll wait.

OK, so now, let's imagine you could travel back in time to experience some of the periods in our planet's life which would have been disastrous for human beings.

About four billion years ago: hell on earth

The Earth was newly formed and our atmosphere was composed mainly of hydrogen and helium. You wouldn't have had time to suffocate, though, because the planet's surface was still roasting hot and you'd have cooked through before you ran out of oxygen. Still, the site had 'great potential', as estate agents say about wrecked houses. No need to stick around here, though, so we'll move on.

About two and a half billion years ago: a useful waste product

The Earth was cooling down, the oceans had formed, but there were still lots of volcanoes erupting, spewing stuff up from inside the planet. Because things were a bit cooler, you wouldn't roast instantly; you'd probably suffocate first. The atmosphere, which had become mostly methane (an ingredient in the natural gas we use today for central heating and cooking), contained more carbon dioxide now (CO_2 for short), mixed with ammonia and a bunch of other gases like nitrogen. Up until this period, there'd have been no oxygen in that air on Earth for us to breathe, and yet progress was being made ... really, really slowly. A very simple form of life had appeared in the oceans, which was able to live in this toxic environment. The oldest fossils of single-celled organisms date back to three and a half billion years ago.

Now, you may consider yourself utterly superior to bacteria – and you'd have good reason to – but despite having no arms, legs or brains, these ones ruled the Earth for about a billion years. (Compared to their reign, we've only been in charge for a blink of an eye.) Little did they know, however, that their slimy empire was coming to an end. Other single-celled life forms called cyanobacteria had developed the ability to take energy from the sun and use it to feed on carbon dioxide in the water. This process is called photosynthesis.

We owe our existence to these humble heroes, the cyanobacteria. Being able to harness solar power was

like having a turbo-charger. It put fire in their veins – or it would have, if they'd had veins. Compared to the other micro-organisms, they were like super-powered mutants. Their waste product, the stuff they farted out, was oxygen. Over half the lifetime of the Earth, they managed to fart out so much oxygen that *they changed the atmosphere.*

Oxygen was toxic to a lot of the other microbes at that time – imagine slowly suffocating in another creature's farts – so, ironically, the creation of the oxygen that we depend on for our survival ended up wiping out a huge amount of the life on Earth. Some early organisms, however, had adapted to using this waste product to create energy themselves. Life was all about recycling, even then. With their competitors knocked out, these new oxygen-breathing forms of life started taking over the planet.

So you see, oxygen-breathers have been winners, right from the start. As well as being incredibly, astonishingly lucky to be the ones who got to keep living.

Half a billion years ago: a world of creepy-crawlies

By now there was plenty of oxygen, so you wouldn't suffocate, and yet the world still wouldn't be one that you'd recognise. The first multi-cellular life had appeared, gradually growing more complex; there were sponges, jellyfish, sea anemones and then trilobites and primitive fish. The oceans swarmed with life, and plants had appeared on land. Distant though this world might seem, these early organisms have a direct connection with your life today, which we'll get to later.

From that point on, there followed millions of years of living and dying, with countless generations of dead matter settling onto the land and the floors of the oceans. The Earth's climate changed again a bunch of times. It was still a work in progress, but it was starting to look a lot more like home.

The Earth's cool jacket

The oxygen that makes our air breathable was only one step towards creating the mighty, planet-altering species that we are today. Even as this was happening, something else was helping destroy the old rulers of the world. You see, because the Earth was a big, spinning ball of rock in space, orbiting the sun, this layer of mixed gases served another vital purpose.

If our world was exposed to the sun without the shield of its atmosphere, the full blast of the sun's radiation would scorch the planet's surface. There would be no animals, no trees, no plant life of any kind, no soil. Most of the oceans would boil off and evaporate into space. Meanwhile, any part *not* facing the sun would be freezing cold, like anything else left out in space without sunlight. That's how we'd be as the world spun, taking turns being cooked by radiation and frozen solid.

It's easy to forget that the Earth is mostly just a giant ball of stone; it's the thin layer of gas around the outside that makes it special. If you want to picture what our planet would look like without an atmosphere, look at the moon.

The atmosphere is awesome. It's like a light jacket that can keep you warm or stop you from getting sunburned. It insulates our world, absorbing or reflecting the radiation and spreading it around so that no one place is taking the full hit from the sun; nor is any part of the planet enduring the killing cold of space.

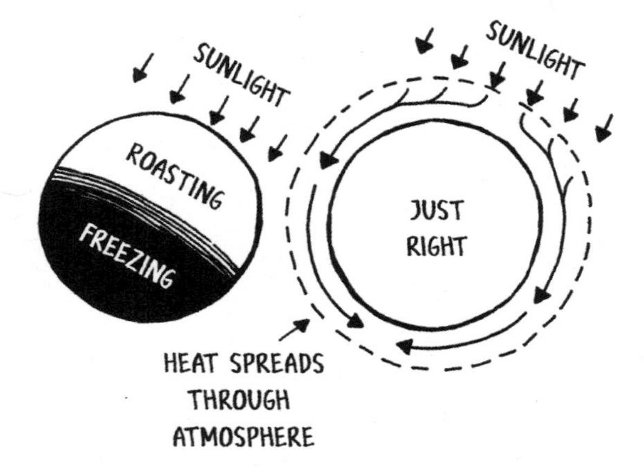

This is what's known as the 'greenhouse effect'. While other planets in our solar system have atmospheres too, they don't support life – as far as we know. Through some extraordinary luck, Earth ended up in what's known as the 'Goldilocks zone', the distance from the sun which let us hit just the right balance of temperature to allow liquid water, which enabled our atmosphere and life to develop. A bonus feature is the atmosphere's ability to deflect or break up smaller meteors, which is pretty cool. Not the bigger ones, unfortunately – they still hit us from time to time.

Back when our bacterial ancestors were stomping their rivals in the early days of oxygen, the atmosphere still had a whole load of methane mixed in. Certain gases in our atmosphere are really good at trapping heat. They're known as greenhouse gases, and methane (or CH_4) is the Incredible Hulk of greenhouse gases. It held onto a *lot* of heat, keeping the Earth very warm. However, like the Hulk, it doesn't stick around for very long. It cycles out of the atmosphere much faster than carbon dioxide. Carbon dioxide is a lot weaker, but it's the clingy, annoying visitor who insists on hanging about after everyone else has gone home.

As oxygen levels rose, way back in the olden days, levels of methane in the air were reduced. At the same time, those busy little cyanobacteria were gobbling up that other greenhouse gas, carbon dioxide. The atmosphere was still acting as a shield, but greenhouse gases were disappearing, so it was holding in far less heat. This, combined with the fact that the sun back then was only a young thing, and wasn't as hot or as steady as it is now, meant that early life was about to hit another challenge.

This reduction in greenhouse gases is believed to be *one* of the reasons (volcanoes were taking a bit of a break too) that the Earth's temperature plummeted about 600 to 700 million years ago, creating a severe ice age. Enormous glaciers, plains of solid ice, crept across the entire surface of the planet, grinding the land beneath them and closing over the sea. It was a situation known as 'Snowball Earth', which is an actual term scientists use. It put an end to untold numbers of poor, innocent microbes.

Again, not a great time to be around.

But some of those tough little organisms survived, bless 'em. The Earth slowly warmed back up as the volcanoes got busy again, the glaciers melted and the atmosphere found a happy medium: not too hot and not too cold – like when you're trying to get the temperature right when you're in the shower. Life crept back, it breathed and grew and spread and became ever more complex. Simple animals began to appear. Algae and wide varieties of phytoplankton joined the cyanobacteria in the seas, creating a steady supply of oxygen. This was increased further as simple plants evolved on land, developing into trees and then into huge forests.

Life, the atmosphere and the oceans all became vital parts of one beautiful, giant, moving system, constantly affecting each other. And it has been that way ever since.

German scientists recently discovered a new strain of bacteria known as pseudomonas bacteria *that eats plastic waste. Its favourite is a notoriously hard to recycle plastic. These bacteria could really help with our waste crisis!*

CHAPTER 2
THEN, ALONG CAME THE HUMANS

Weird as it may seem, you have something in common with those bacteria that changed the climate billions of years ago. Your body produces hazardous substances – hazardous even for you, which seems a bit mad, but that's just how Mother Nature is; she giveth and she taketh away. These substances also affect the world around you.

If you decided to do absolutely nothing tomorrow but lie on the couch and snooze, you would still be changing the chemical composition of the air, by taking in oxygen and exhaling carbon dioxide. The same is true for every other animal on Earth. Billions upon billions of us are at it, breathing, every second of every day. Now, if there's too much carbon dioxide in the air, it can poison you. Our air is about 0.04 per cent carbon dioxide. If it went up to 10 per cent, it would be too toxic to breathe. We can die of our own waste product.

Fortunately, this doesn't happen, because the Earth has a way of balancing out our exhalations. It has masses of plant life, which soaks up carbon dioxide and gives off *oxygen*. As long as those plants are giving off more oxygen than animals take in, we get to keep breathing. What does a tree do with all this carbon dioxide? you may ask. Well, it uses sunlight to convert CO_2 and water into glucose and oxygen. The glucose is its food, and it uses the carbon in it to build. Think of the biggest, oldest, strongest tree you know. Most of that bulk didn't come from the ground – otherwise there'd be a big hole around the tree – it came from the *air*.

Trees are built mainly out of carbon drawn from the air, and *their* waste product keeps us alive.

For obvious reasons, that balance is not something we want to mess with. But long before we interfere with our own breathing, it's possible to interfere with the atmosphere's other role: the protective jacket it forms around the Earth.

Real-life magic

Now, let's look back again – this time, just a few hundred thousand years ago. Life on Earth had changed utterly from when we'd last visited it, half a billion years before. It had been a rough ride over that time, including a bunch of other ice ages and mass extinctions that did away with some pretty cool creatures like the trilobites, the dinosaurs and the giant shark, the megalodon – which is probably just as well, to be honest. I mean, who needs that kind of competition?

Now, an *ape* that walked upright on two feet was starting to make its mark on the Earth. It wasn't the strongest creature out there, or the fastest, and it didn't have lethal teeth or claws. In fact, compared to some of the other animals around at the time, it was pretty pathetic. But it was smart. It was already using tools and co-operating in tribal groups. It was developing language, so that each generation could pass on knowledge to the next. These apes were improving themselves in a way that sent them racing ahead of evolution. They were working on a different time-scale to the rest of the planet – and they were speeding up. This species of ape could take on much more powerful animals by killing them with rocks, and then sticks, and then sharper sticks, and then it learned to throw the sharp sticks and then ... well, you get the idea. Humans had arrived.

And somewhere, at some point all those hundreds of thousands of years ago, they learned to make fire.

Early humans would have been *using* fire for a long time before that, finding burning material left over from lightning strikes and forest fires, but now they had worked out how to *make* it for themselves, any time they wanted, by rubbing two pieces of wood together. Fire was one of the most destructive forces in nature, a monster that every other animal was terrified of. But humans loved that little fiend. And once they'd domesticated the monster, they started using it in as many different ways as possible. Suddenly, no other creature could compete with these lanky apes. Fire was *magic*.

Fire kept humans warm, allowing them to spread out and adapt to different environments. It provided light for them to see at night, which meant they could hunt and work for longer. It kept predators away while they slept. It enabled them to make new tools. Grass fires could be used to drive prey in a particular direction during a hunt. The humans learned to cook their food, which killed off bacteria and parasites and made it easier to eat. Biting, chewing and digesting took less energy and it meant that foods that were too tough to be eaten before could now be consumed with much greater ease. Humans lived longer, they increased in number. They learned more with each passing generation.

Fire raised them above the rest of the natural world.

Despite all the smoke they were releasing into the air, these little fires the humans were making were not a problem. Fire, a monster though it might be, was already a natural phenomenon, part of the cycle of life, and humans' small contribution was hardly noticed among the seasonal forest fires and the volcanoes. Everything was just *fine*.

The invention of spare time

It wasn't long before humans were dividing their environment into three categories: 'Can we *eat* it?' they would ask. 'Can we *make* anything out of it?' 'Or can we *burn* it?' When it came to many types of plants and animals, the answer was 'Yes' to all three questions. The humans conquered all before them. If some aspect of their environment could not be beaten or tamed, they just set their bright little magic monster on it.

The more control they had over their environment, the more effective they were in feeding themselves. And as they developed and gradually extended their lives, humans discovered another miracle: spare time. They had moved beyond merely staying alive; now they could take an occasional break and sit around and chat. They had time to ponder over problems, experiment and take risks that they couldn't before. These were the first hints of *science*. They began to communicate by visualising ideas with paint on the walls of caves and with carvings in wood and stone – to make *art*. They created religions to help make sense of the world, and to help define what they were. The different members of the tribes could specialise, devoting themselves to hunting or gathering, cooking or making tools as well as raising and teaching the young. They could look after the old and sick members of their tribes, preserving knowledge. They could make plans for the future.

Food that obeys

And that was how our single greatest development after the control of fire came about: the beginning of agriculture. Around ten or fifteen thousand years BCE, humans decided they'd had enough of chasing their food all over the place, and that it might be a better idea if the food just stayed put and did what it was told.

Instead of spending all their time hunting wild animals, they started breeding tame ones that didn't run away when the humans wanted to eat them, starting with sheep, pigs and then cattle. However, the tribes of humans were still wandering around as this went on, their meat walking with them as they followed nature's cycles, harvesting what plants they could from one location and then moving on to another. They never stayed anywhere for long, so they never had a major effect on any one place.

Then they came to understand the cycle of seed to mature plant, and began to sow simple crops like wheat, barley and rice. And the wandering stopped. Unlike meat, growing crops can't walk with you until they're ready to be eaten, and you certainly don't want to go off and leave them for someone else to find and eat. Humans had found a way to ensure their environment provided them with food, food that grew in the quantities they needed, but the price was that they had to settle on the land, and their animals had to stay with them. And now that they were stuck in one place, the humans took to changing their environment instead of adapting to it. This was the beginning of civilisation.

And it was at this point that Mother Nature lifted her head suspiciously and sniffed the air.

Friends of the Earth Ireland

Art can be a great way to get people thinking about global issues and inspiring change. 'Craftism' is where craft meets activism. Human creativity is a key ingredient in changing the world for the better. Let's nurture creativity!

CHAPTER 3
OUR ANIMAL NEEDS

Civilisation has given us so much, it's easy to forget that we are still animals. We need to breathe, drink water and eat. We need physical comfort, safety and shelter. We need to reproduce and raise our young. We need a safe way to dispose of our bodily waste. These have always been our most important challenges, and they were the first problems that civilisation tackled and learned how to solve, so now most of us tend to take the solutions for granted. If you want a drink of water, you can turn on a tap. Food comes to a shop near you; you don't have to go and harvest it or hunt and kill it yourself. Need to pee or poo? There's a flush toilet for that.

However, these extraordinary benefits came at a price. We had to burn a *mind-boggling* amount of the Earth's surface to get where we are today.

We need fuel and oxygen to make fire. All the things we burn on a large scale for fuel – wood, turf, coal, gas, oil – started off as living matter, mostly plant matter. Some fuel can be produced in a few years, like wood, but others, like coal or oil, can take millions of years to form naturally. Living matter contains a lot of carbon, and when it's burned the carbon is released into the air as carbon dioxide. The same farming that enabled us to create civilisation also marked the time when we started setting fire to things in a *big* way. Up until then, we hadn't messed with the natural cycles that created and maintained the connection between us, our atmosphere and the oceans. That was about to change.

Our smoking habit

If there's one activity that truly demonstrates humans' rather short-term attitude to nature, it's slash-and-burn farming.

Slash-and-burn does exactly what it says on the tin. You cut down a section of forest, let the fallen wood and vegetation dry out, and then set fire to it. The ash fertilises the soil and you plant your crops in it. But that fertile earth doesn't last. After a few years, the nutrients in the soil are used up because you're taking away everything that's grown in it, and nothing's going back in. So you have to move on to another section of forest, creating another spread of fields. Used on a large scale, it destroys trees, which never get replaced, and creates huge swathes of land that end up useless for farming and not much good for wild vegetation either – or the creatures who depend on it. Not only were we destroying trees, but we were also killing *much of the life that lived in the trees*, everything from insects to large mammals, including other apes.

Back in the early days of farming, we could at least

blame this attitude on pure ignorance. We just didn't know any better. The world felt enormous, and it was inconceivable that mere *animals* could ever do anything to have a permanent effect on it. Even when we started cutting down large areas of forest to make way for farmland, it was slow and painstaking, and felt like nothing more than clearing a dot on a vast landscape.

Slash-and-burn can be made to work on a small scale, when used responsibly, where fields are left to recover after harvests; but the widespread burning of trees was the first wound we cut into the Earth. As well as the destruction it caused on the surface, slash-and-burn released masses of carbon dioxide into the air, and because there were no new trees planted to balance it out by absorbing carbon *from* the air, we were, for the first time, interfering with the complex cycle of life, earth and ocean that maintained our atmosphere. And though these were still only small injuries, ten thousand years ago, we were already starting to damage our planet faster than it could heal.

And the smarter we got, the more damage we did.

Warping time

Ten or fifteen thousand years feels like a very long time in human terms. We've certainly got a lot done over that period. To a planet a few billion years old, however, it's a brief moment. Imagine going to bed with a full head of hair, and then waking in the morning to discover half of it has fallen out because of some hyperactive parasites on your scalp. Imagine it keeps falling out through the rest of the day, until there's none left; you're completely bald by the time evening falls. Every time you blink, the parasites have changed something about you.

That's what we're doing to our planet. For us, it has been thousands of years of building a civilisation; for the Earth, it's been one really bad day.

Humans were lucky we came along when we did. We could not survive on Earth today without the billions of years' worth of life that came before us. That's a *lot* of life that had to happen first. As organisms in the sea, as well as the forests and other vegetation, lived and died and rotted and gave themselves to new life, layer after layer over thousands of centuries, this dead matter created the soil we rely on for growing our food. This early plant life also clung onto much of the carbon it had absorbed from the air, holding it close as it settled over time into deep layers across the earth.

As our hunger for fuel grew, we found new kinds, some of which burned longer and gave off more heat than wood. Peat (which we often call turf in Ireland) formed in wetlands over thousands of years. Dried out, it was a potent, convenient fuel, and anyone who knew how could just dig it out and cut it up. If you were lucky enough to live near a bog, you could literally burn the ground. Even as forests were thinning out, as they were cut down for fuel and building materials, the peat bogs continued to provide.

Peatlands are one of the most effective environments for absorbing carbon. The water-logged conditions slow the decomposition of vegetation as it dies, preventing it from releasing the carbon dioxide that it has absorbed.

In Europe alone, peat bogs lock up about five times more carbon than forests. Unfortunately, it would take the lifetime of a peat bog, and the destruction of many of them, for us to understand why peatlands are so crucial to our environment, and to start making moves to protect them.

As working metal became more important for making tools, weapons and eventually machinery, we needed much hotter fires to melt and cast it. And as luck would have it, the Earth gave us coal. Huge reservoirs of energy, held in hard black slabs of carbon, begging to burn. Coal was what peat became when it grew up. Pushed further and further down into the earth by the layers forming on top, it had been crushed flat, subjected to huge pressure and heat deep underground for millions of years. And all the carbon that peat held was concentrated down, compacted so tightly it turned to a sheet of black rock. It was capable of giving off incredibly intense heat when it burned. The first humans to use it must have felt like they'd discovered a new, even more fiery type of fire. A gift from the gods.

With fire like this, we could finally master metal. Our civilisation had advanced another step.

A monstrous appetite

Thousands of years passed after those first early steps into farming, and humans were learning the whole time, developing, speeding up, our progress unstoppable. We spread all over the world, and wherever we went, we cut into the planet's surface and took whatever we could find. The 18th century saw the birth of the Industrial Revolution in Europe and the United States; the first large-scale factories were appearing, and steam-driven machines were increasing the amount of work we could do every day. We *used* more of everything; we *made* more of everything. Machinery was starting to replace workers – and much of the power for that machinery came from coal. By the year 1750, Britain, then the most industrialised nation in the world, was producing 5.2 million tonnes of coal *per year*. And that was just one country.

We were pumping millions of years of stored, concentrated carbon straight out into our atmosphere every day.

In the mid-19th century, another type of fuel, and not a fossil fuel this time, was also proving invaluable to our new industrialised world: whale oil. Highly prized for use in lamps and for lubricating machinery, it was in demand across Europe and North America, and came to be used for anything from margarine to car gearboxes. This type of oil was not doing much harm to the atmosphere compared to the burnt dirt of coal, but it was wiping out enormous

numbers of whales, and damaging the ocean ecology of which they were a valuable part. A huge industry built up around the hunting of whales, and it was so successful that by the early 20th century it came close to driving some species to extinction.

Driving species to extinction is something we've become particularly good at.

It's ironic, perhaps, that the main thing that saved those endangered species of whales also helped to speed up the growth of our civilisation even further, and in doing so, ensured the rise of a much greater environmental crisis. Humans discovered petroleum. Here was a highly efficient fuel that burned like coal, but came in liquid form, like whale oil. It could be used for a wide range of purposes – and you didn't have to chip it from the wall of a dangerous mine with a pickaxe, or chase it across the ocean to catch it. It was lying right there under the ground; you just had to drill down and pump it out.

We'd struck oil.

Friends of the Earth Ireland

Natural forests and healthy peatlands absorb carbon from the atmosphere. They are also havens for nature and wildlife. Native trees planted in appropriate areas help us fight climate change. As for bogs, Scotland and Indonesia are leading the charge with ambitious projects to bring life back into these often damaged and forgotten landscapes.

CHAPTER 4
A PLANET-SIZED FUEL TANK

Petroleum, or what we commonly just call 'oil', as well as all the things that are made from it, is so woven into our lives that it would take a much larger and more complicated book to describe that relationship properly. Having left behind the stone age, the bronze age and the iron age, we are now living in the oil age. No one substance has had a greater influence on our modern

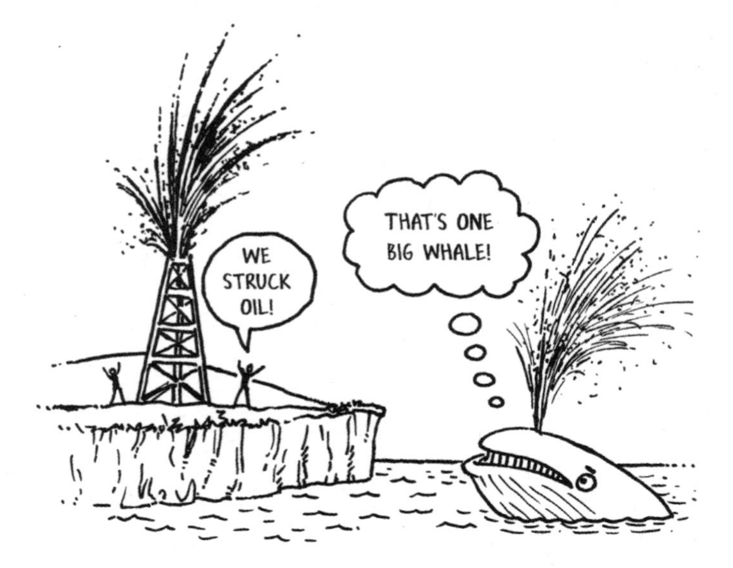

civilisation than oil. Apart from all the forms of it that are burned for energy, it's also used for everything from crayons to car tyres, from drugs to toilet seats, from fertiliser to lipstick. And it is the source of much of our plastic. Right at this moment, it's likely that you're wearing something with an oil product in it, or sitting on one or holding one – or all three. It is *everywhere.*

But mostly, we burn it.

While we use oil, coal and natural gas for the manufacture of plastic, a material that plays a part in every aspect of our lives, the vast majority of these three substances are burned as fuel, either for transport or to generate the electricity we rely on for ... well ... almost everything. It would be hard to imagine life without them.

But our old friend fire, who has been with us from the start, has been working some dark magic on us. Fossil fuels are hydro-carbons – compounds formed by combining carbon and hydrogen. When they are burned in the presence of oxygen, the chemical reaction releases carbon dioxide into the atmosphere, which starts wielding its quiet influence. CO_2 is the most important gas for controlling the Earth's temperature, not because it can hold onto the most heat, but because it's just always there. It *stays* carbon dioxide, while other gases react with each other, change and are cycled out of the air. Carbon dioxide is the persistent hanger-on, and its concentration affects water vapour and clouds, which have a bigger impact on temperature.

And though nature already releases enormous amounts of CO_2 – many times more than we do – the world's life cycles have managed to maintain a balance that keeps our atmosphere at that Goldilocks level of 'not too hot' and 'not too cold'. Now that we're adding to it on a big enough scale, we've started to knock things off balance. We're spewing it out faster than Mother Nature can draw it back in again. Slowly, ever so slowly, the heat is building up in our air.

Even if you have no interest at all in protecting the environment, *imagining a life without fossil fuels* is exactly what we have to start doing. The amount of oil left in the world – at least, the stuff that it's possible to reach with existing technology and without devastating environmental impacts – will last us less than a century at the rate we're using it. And since we first started burning it, our appetite for oil has steadily increased.

For a long time, we've worked on the basis that we have a planet-sized fuel tank, vast reservoirs of energy that are just ours for the taking. Unfortunately, that tank didn't come with a fuel gauge, so we didn't worry too much about how much was in there and maybe try planning for the future a bit, given that we had no way of refilling it.

It is extraordinary to think about it, but just a few generations of humans, and mainly the wealthiest humans, will have used most of the Earth's reserves of oil, coal and gas – reserves that took millions of years to form. We will have left very little for those who come after us.

Our civilisation was built on these fuels. The story of our development is a story of fossil fuels, and we have to accept that fact, even as we accept that our world has paid a heavy price for the dizzy heights we've achieved. And now we've come to realise that we simply can't keep burning those reserves until they run out, because the changes that would cause to our climate would threaten our very existence. Our past was built on fossil fuels, but they have no future. Even the oil companies are realising it. You can tell, because they're all starting to invest in renewable energy. *That's* where the future lies. The same progress that has driven us to this point is carrying us onwards.

When the first astronauts looked down at the Earth, seeing this shimmering blue ball hanging there in space, they experienced what's known as 'the overview effect', a profound awe of the astonishingly rare, delicate and beautiful world we live on. And yet they had been carried up there by massive rockets burning hundreds of thousands of litres of fuel. We are the first species on Earth to reach into space, our incredible advances giving us a unique understanding of our environment,

but those same advances have come close to *destroying* that very environment.

We have gained this insight just in time to do something about it. Now it's time to take the foot off the gas. Our story has to start taking a different path.

Eco-anxiety

Seeing the challenges our world faces, and understanding their significance, has an effect on us. It is common for people to experience anxiety – often referred to as 'eco-anxiety' – as a result, and this is a perfectly natural reaction. It shows an awareness of and an empathy for the wider world and a willingness to accept painful truths. It is a completely normal, healthy response to feel fear, anger or sadness in the face of a real threat. It's your body switching to emergency mode, even when the emergency is so long-term there's nothing you can do about the problem here and now, which can leave you with no tangible way to channel these feelings. It's important to acknowledge this, and to take care of your mental health. It's vital, too, that you share your feelings with others, to talk about it; and being part of a community of like-minded people can be a big help. Many people find that taking action in some form goes a long way to channelling and easing that anxiety.

If you are feeling overwhelmed, if it becomes unbearable, or if you find yourself unable to function normally as a result of your feelings, then you might need to seek professional help. This is a troubling subject, and it has an effect on us that must not be underestimated or ignored. As with any subject covered in this book, we have to rely on the experts to help us tackle the problems we face.

Normality glasses

Anxiety can be caused by awareness, but there's a flipside to that – an *unconscious resistance* to awareness. I want to talk about the here and now, and the hold 'normal' has over us. We often look at the world through normality glasses, and by that I mean our firm conviction that normality doesn't really change as time passes. It's human nature to think that what is normal now will always be normal.

When I think of the future, I make assumptions based on the life I have now. There are so many things that I assume will stay pretty much the same for years to come – and this is despite the fact that experience has taught me that *it isn't true.* Our normality is very different now to the one I was born into. I was born in Ireland in 1973, and here are some of the things that were normal at different stages of my life as I grew up, but have now changed completely:

- Nobody had mobile phones. You had a phone in your house, and it remained attached to the wall.
- Children were still beaten in school. Corporal punishment was only banned in 1982, and it took some teachers a while to get out of the habit.

- Cities in Ireland and the UK were sometimes choked by the smoke from tens of thousands of coal fires, and the sky was sometimes tinged brown. Smoky coal was banned in Ireland in 1990.
- Computer games were just starting out and were *really* basic. Hardly anyone had personal computers in their homes. The worldwide web did not exist until 1993, when I was twenty years old. It still took years before people were able to use it in homes and businesses.
- Teachers' staffrooms smelled of cigarette smoke. So did most rooms in most buildings. Smoking in the workplace was only banned in 2004.
- There were constant shootings and bombings in Northern Ireland. Between 1969 and 2007, British troops patrolled the streets, and in some areas, the police had to travel around in armoured cars. If you crossed the border into Northern Ireland, you could be searched or questioned by heavily armed soldiers.
- We had roller discos, where you went to roller-skate in circles to music. This was a real thing.
- Oh, and there was the Cold War. The Soviet Union and the United States were constantly threatening each other with nuclear weapons. In the 1970s and 1980s, we lived with the very real possibility that someone might start a nuclear war that could *wipe out all human life on Earth*. So there was that.

Despite what you might think from all this, I had a mostly happy childhood. For anyone living back then, these things were normal. And yet, imagine if troops appeared back on the streets in Northern Ireland, or nobody had mobile phones, or a teenager's night on the town involved roller skates, or you could look up in the evening and see a brown sky. Imagine if teachers started hitting children again. Imagine living in fear of nuclear war. That would be a fundamental change to what you'd consider normal.

This kind of major change in what we think of as normal is something you've already lived through. In 2019, the COVID-19 virus started to spread across the world. Society went through some extraordinary changes. Schools and businesses were ordered to close for months. People couldn't meet each other or visit their loved ones in hospital. There were restrictions on travel. People had to wear masks when they went out in public. All over the world medical researchers began desperately developing vaccines and treatments. At the time of writing, the world was still trying to adapt to this upheaval.

Normality changed.

And normality is going to keep changing, in positive as well as negative ways. We know this. We will have advances and capabilities in the future that we aren't even thinking about now. And there will be things that we have now that we'll have to give up, because either they are going to run out, or because using them is putting our future in danger. The only thing we can say for certain about our future is that our world will be *different*, and we won't have control over some of the biggest changes we're facing. But the more we're ready to accept change and take action to deal with it, the more control we will have.

We can't wait and see

There is a phenomenon in any disaster known as 'normality bias'. It's when people are faced with some major threat, and they suffer a kind of brain freeze. It is an extremely common reaction. Because most things don't change much from day to day, we get used to that normality, and assume that it will always be the case. And when faced with a sudden threat to your life, it can take time and effort to change gear into full-on emergency mode. After all, you've got through every day up to this point in your life, so you think surely you'll survive this one too. Normality bias happens in everything from serious illness diagnoses to car crashes, from swimming accidents to natural disasters. Dealing with a life-threatening challenge can be overwhelming, making massive demands on your body and mind. There's a real temptation not to act, but to just hold on, and wait and see if everything will be all right.

And while there are millions of people all over the world taking action to fight climate change, and there have been for years, the human race as a species is hesitating in those few vital moments before disaster strikes. Our normality is changing, and we can take control of that change. Or we can wait and see what it does to us – not such a great option.

But this thing doesn't know who it's messing with. We're the species that's been using intelligence and ingenuity to overcome everything the world has thrown at us for tens of thousands of years. We're the species that mastered fire and invented farming.

We've explored the bottom of the ocean and we've reached into space.

We've never taken on anything on this scale, but we are ready for this. Tens of thousands of years of solving problems has *made* us ready. That said, it's so big that every one of us has a part to play – we have hard decisions to make, and we need to make them fast.

And to deal with the challenge we're facing, we need to understand it.

Let's take off those normality glasses and time-travel 100 years into the future. Imagine: the world is safe for all people and nature is thriving again. What does society look like at this time? How did we get there? Use your imagination to envision how we want the world to be.

CHAPTER 5
UNDERSTANDING THE CHAOS

Weather. It is a constant source of conversation, and affects everything from our food to our oceans, from our travel to our emotions. For most of the time we've been in this world, humans have had some sense that our weather had patterns, but most of us had little understanding of them, what caused them or how to predict them. A lot has changed over the last few hundred years, and now meteorologists have attained a knowledge of the weather that people a few generations ago could only have dreamed of. With a vast range of instruments at their disposal, including satellites and computer modelling, scientists can observe weather, study it, analyse it, describe it in detail, and even predict it.

I am not a meteorologist, so I'll be keeping this *really* simple, which – let's be honest here – is about my level.

The power source – how weather happens

The Earth is a spinning ball, so although the part of it that is exposed to the sun is always hotter than the rest, that heat is constantly moving across the planet's surface. This is where the jacket of gas wrapped around the globe works its magic. Instead of just heating that one patch of rock and then leaving it to freeze as it turns away, the sun's rays heat the Earth's surface and some of that heat is radiated back out and absorbed by the greenhouse gases in the air, such as carbon dioxide (CO_2), methane (CH_4) and nitrous oxide (N_2O). These gases, combined with water vapour in the warm air, hold onto that heat and then disperse it again over time. This is the greenhouse effect.

That air *moves*, carrying the heat with it. The warm part expands and rises, going all light and fluffy, while the cooler air around it gets all heavy and sulky. Because this cold air is thicker, denser, than the warm air, it has a higher pressure, so it pushes in under the warmer stuff and sets everything swirling, which becomes even more chaotic as it interacts with the features on the Earth's surface.

If you've ever watched a pot of water come to the boil, it's a similar effect, though more concentrated. As the water heats, the hotter stuff turns to water vapour, forms bubbles and rises to the top. If you turn the hob down and leave the water to simmer, you'll see it's still constantly moving. Our atmosphere is doing a much bigger version of the same thing. It's just not as hot – and it doesn't make bubbles.

So the heat spreads across that sunlit area of the surface, but the Earth is always turning, the sunshine sliding westward. The warm air mixes with colder air around it. With this mixing movement, and the changes in pressure, the air currents take the intensity out of the sun's heat by sharing it around the planet. This keeps our whole world at a nice in-between temperature, instead of being a dead rock that's roasting on one side and freezing on the other. And that churning of hot and cold air and changing concentrations of water vapour are the main cause of our winds – and the rest of our weather.

The oceans are warmed by the sun too, and that change of temperature causes the currents in water, just as heated air causes winds; but the change is much slower and longer lasting. The oceans can absorb a *lot* of heat. That's something we'll come back to later. The main thing to remember here is that it's the sun's radiation, absorbed by our atmosphere and oceans as they spin around, that powers our weather. And one happy consequence is that this process *helps stop the sun from killing us all.*

But it's a very delicate balance, and the more heat our planet absorbs, the more powerful our weather becomes.

Climate is not the same as weather

The planet is always spinning in the same direction. The way the sun heats it does change with the Earth's orbit – but only a little and very slowly, over time (at least, from our point of view). Because the cycles of heat and rotation are consistent, the major winds and ocean currents flow in regular patterns that meteorologists have come to understand. Moving like some impossibly complex clock, all the tiny little weather events happening around the globe can seem utterly random, and yet there is a system to it all. Everything is connected. And one key to understanding it is to know that there's a difference between climate and weather.

Weather changes from day to day – or in Ireland and Britain, from minute to minute – while *climate* is how all that weather averages out over a long time. We can have days as hot as the south of Spain, or as cold as Norway. However, while days like these might be ordinary for these countries, for us they are just short swings to the extremes – neither example would be typical of our climate. This is why someone can't wave at one day of very cold weather as proof that global warming is a hoax. In Britain and Ireland, our climate is a moderate one: not very hot, not very cold. We can have short periods of heavy rainfall like India's monsoon, or dry spells like Afghanistan, but they never last long. All of these countries have different climates to ours;

extremes that are normal for them are unusual for us. One of the features of the climate in Ireland and Britain is that our weather is very changeable – and therefore hard to predict. And all these climates are caused by each country's place on the globe, as well as their unique geography and the particular winds and ocean currents around them.

It's like the difference between your character and your behaviour. You might act the fool every now and again – most people do – but most of the time, you are *not* a fool. Your character is who you are all the time. Your behaviour is what happens moment to moment, and it's influenced by your character, by who you are.

Climate is like *character*, and weather is like *behaviour*.

Small numbers, big changes

When scientists tell us our climate is changing, they're not talking about the weather in any one place, or what it's like over a few days. They're not even talking about the climate in one region, or over a year or two. They're talking about the climate of the entire Earth, and changes that are measured over decades. The Earth's average surface temperature is about 14°C, but they've had to average this out over a lot of figures because there's so much to measure – and different ways of measuring it – that it would hurt your brain. (Well, it would hurt *my* brain.) Really smart people spend their whole lives just studying small parts of this stuff, like the bubbles in ice cores drilled in the Arctic, or cores of wood pulled from ancient trees, showing the years of rings.

Gathering huge amounts of information over years, climate scientists can see the patterns of the weather and put them together to get the global picture, like piecing together the world's most insane jigsaw. When they have enough of the pieces, they can stand back and see the whole thing, and a picture of our climate's history is formed.

While weather might change by ten or twenty degrees in one place over a short period of time, making it hard to predict, the *climate* of a much larger area tends to swing back and forth much less. By the time you get to the scale of the whole Earth's climate, it might have lots of little things happening all over it, but the big stuff runs like a clock, regulated by the spin of the planet and powered by the sun. It takes something pretty extraordinary to change anything on that scale. It's hard to imagine how much energy it would take to raise the entire planet's temperature by even one degree Celsius.

So how could you possibly make a whole planet warmer? Well, you'd either have to increase the amount of heat the sun blasts at us, which would mean turning the sun up – not something we could ever do, or want to do. Or you could move the Earth closer to the sun than its current orbit allows, which would also be pretty impossible. Or you could increase the planet's capacity to hold onto that heat. Which is what we've done. We've thickened the lining of the Earth's jacket. We've changed the composition of our atmosphere enough

that it's holding onto just a bit more heat than it did a hundred years ago. And that heat is *added energy*. In the last century, we've turned up the power on our weather.

Friends of the Earth Ireland

In 1859 Irish physician John Tyndall was the first person to prove the link between warming air and atmospheric concentrations of CO_2. Today we call this the greenhouse effect. His studies took him to the Alps, where he also became an excellent mountain climber.

CHAPTER 6
MAYBE IT'S NOT JUST US?

The Earth's climate doesn't need us to be weird; it can do that all on its own. People who argue that climate change has nothing to do with humans will often raise this point to prove their case. The fossil-fuel industry has spent fortunes on trying to discredit or contradict the science, confusing the issues and down-playing the threat, and they have been very successful at it. Climate change deniers will say that many different things have affected the world's climate over the four and a half billion years it's taken to get to where we are now. And that much is true. Some were one-off events; others are happening all the time, in very, very long cycles. Let's have a look at the main culprits.

The sun

The sun has grown stronger over its lifetime, so the amount of radiation that hits us now is greater than the amount experienced in the early years of the Earth, but this is a change that can only be measured over billions of years. As well as this, our orbit around the sun isn't a perfect circle – it's an ellipse. And it has also shifted continuously over the life of the planet. This means that there are times when we are further from the sun, or closer to it. For obvious reasons, this can cause the Earth to get colder or warmer. The effects of this occur over thousands or tens of thousands of years. Powerful bursts of activity like solar flares can affect the planet too, particularly our magnetic field, messing with power grids and electronics. These events are, by their nature, very short-lived, lasting only hours or days.

The angle of the Earth

The Earth doesn't sit perfectly vertical in relation to its orbit around the sun, with the poles pointing straight up and down. Like the globes you see everywhere, it's slightly tilted. Our orbit means that sometimes the North Pole is tilting away from the sun, and other times it's tilting towards it. The angle of this tilt also varies between 22° and 24.5°, but it happens over a period of about 40,000 years, so it's not like we'd notice the world tipping over. A change in angle means that the sun's radiation is focused on a slightly different area of the planet; when the Earth tips over further, summers become warmer and winters become colder.

Meteor impacts

Large meteors striking the surface of the planet, like the one that helped wipe out the dinosaurs, can send mountain-sized clouds of dust into the atmosphere, which can block sunlight and cause a drop in temperature. Depending on the size of the impact, this effect can last for years. Thankfully, it's been a while since this has happened.

Volcanoes

Like meteor impacts, volcanoes can blast masses of dust into the atmosphere, causing global cooling, but they can also release carbon dioxide and water vapour, increasing the greenhouse effect of the atmosphere and warming the planet too. This kind of activity used to be a lot more common, but the Earth's crust has chilled out and settled down a bit. The world's volcanoes just don't churn out stuff like they did in the good old days. Even so, they remain a part of the Earth's carbon cycle, and they won't be going away any time soon.

The moon

The moon's orbit causes the tides, as its gravity pulls on the Earth. It can also cause changes in atmospheric pressure, which could affect rainfall, though these effects are pretty minimal. So, the moon doesn't have much influence on our climate, but its orbit can increase the impact of things like storm surges, which cause more destruction at high tide.

Life

At the start of the book, we learned that even humble micro-organisms can change the atmosphere, given enough time – like, millions and millions of years. Other mass developments of life can influence it too. Like us. The greenhouse gases we generate, like carbon dioxide and methane, can turn the Earth into a snowball or a barbecue, depending on their concentration. And we're a lot more powerful than any bacteria.

Even so, how can we be sure that humans are causing this onset of climate change?

The hockey stick

All the factors that can affect the Earth's climate are incredibly complicated, and though it's been a very slow, steep learning curve for scientists, humans have also been studying this stuff for thousands of years. We've been able to take accurate measurements of basic things like air temperature, air pressure, rainfall and wind speeds, and keep records of them, for well over a century. And by cutting into ancient trees and digging into the deep layers of rock, ice, sea sediment and coral reefs, scientists have been able to learn extraordinary amounts about the changes in climate over the lifetime of the Earth. And since it has changed so much in that time, it would be easy to believe that the changes we're seeing now are just another part of that natural rise and fall of global temperatures.

Which is where the hockey stick comes in.

In 1998, a climate scientist named Michael Mann and two of his colleagues put together a load of data that established the average surface temperatures of the Earth going back about five hundred years. Laid out in a graph, it showed that temperatures in the northern hemisphere rose sharply in the last century. For half a millennium, the planet's temperatures went up and down, rising steadily, but very slowly. For most of that time, those temperatures were affected by the kinds of natural factors climate scientists expected to find. But then the temperatures started climbing fast. And they didn't stop.

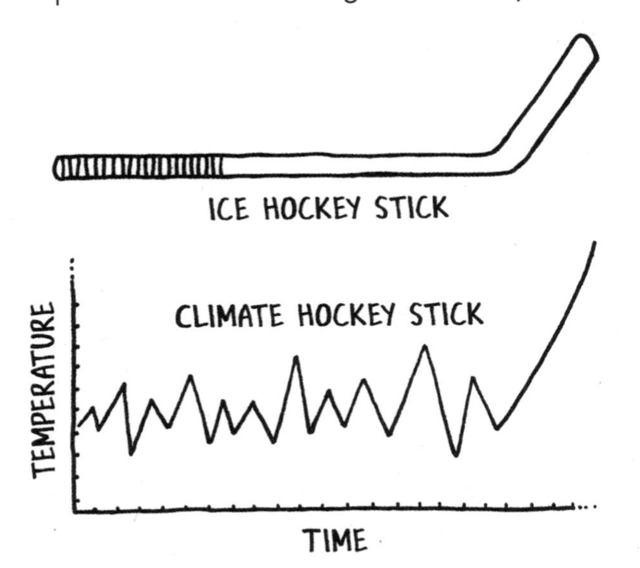

This graph soon became known as the 'hockey stick' because ... well, it looked like an ice-hockey stick. It confirmed what a lot of people had been claiming for years, but there had been no single image to get the idea across that ordinary people could understand. On seeing this data, scientists did what scientists do when they come across something strange: they scienced the hell out of it, prodding and pulling at the data, testing and re-testing, asking awkward questions, and gathering more information.

In case it's not clear yet, I think scientists are cool, and they're going to help us save the world. But let's get back to the story ...

Before long, a whole bunch of other teams had added to, and improved on, the data drawn from things like ice cores and tree rings, and now the 'handle' of the hockey stick was extending back over a thousand years. It was only in the 20th century that they were seeing this shocking rise in global temperatures, and it was the *speed* of that rise that was alarming.

It had taken about five thousand years for the planet to warm by a whopping five degrees as it recovered from the last ice age. That was the difference between the ice age and the climate we think of as normal: just *five degrees*. And yet, in the last hundred years alone, the temperature had risen by more than half a degree, and it was increasing in speed. Based on this curve, the predicted rate of warming for this century was estimated to be twenty times faster than previous centuries. And this was without the kind of catastrophe that normally caused a rapid global change, like a series of massive volcano eruptions or a huge meteor strike.

In 2001, the United Nations' Intergovernmental Panel on Climate Change declared that the increase in temperature in the 20th century was likely to have been the largest of any century during the past one thousand years.

These were the teams of scientists the UN had tasked with gathering all the available research on climate change and putting it through the wringer to hone it down to the best possible data. They were serious people. They were very careful about the things they said, about the language they used. Scientists don't like using dramatic language because they're supposed to leave emotion out of their descriptions. They're supposed to stick to the facts. If the same message had been put out by, say, someone who normally writes fiction, they'd have used much more dramatic language. Maybe some swearing. And exclamation marks.

But scientists do not use exclamation marks. Even so, anyone who understood the significance of this got the shivers looking at that graph. Here was proof that we were suffering global warming at a disastrous rate, without any obvious cause for this disaster.

There was no unusual solar activity to explain it. Volcanoes were not erupting all over the planet. Everything *seemed* to be OK.

People who did not want to believe in the threat of climate change attacked the data, and heaped scorn on Mann and his team, but there was no arguing with the simple facts. Temperature is a pretty basic thing, and there was a range of ways to double-check the facts. The more people questioned it, the more scientists got involved and the more information was discovered, and the further back the handle of that hockey stick went: long, slow rises and falls all the way back to the last ice age – a long, undulating line with one almighty climb in the 20th century.

But what was causing it?

The other hockey stick

When scientists looked at carbon dioxide levels over the same period – something else that could be measured with a high degree of accuracy – they found they had a graph that matched the temperature changes very closely. In fact, levels of carbon dioxide in the atmosphere today are higher than at any time in the last 800,000 years.

Think of it like orange squash, or tea or coffee, whatever beverage you drink. How good it tastes really depends on the concentration of the flavour you're adding. Not enough, and it's weak and watery. Too much, and it's disgusting, overpowering. How much any gas is 'diluted' in the atmosphere is measured in 'parts per million'. During the last ice age, carbon dioxide levels were about 200 parts per million. Before the Industrial Revolution, when we *really* started pumping smoke into the air, they'd risen to 280 parts per million, which was a slightly stronger cup of tea. After a period of thousands of years, this mix of gases was holding in a bit more heat. Between 200 and 280 parts per million is OK. We'd have been fine for them to stay that way.

But in 2013, carbon dioxide levels passed 400 parts per million for the first time in recorded history.

That wasn't good.

The fact that these two graphs matched wasn't enough to prove that the rise in carbon dioxide was causing the rise in temperatures – just because two things look connected doesn't mean that one is causing the other. More sciencing was called for.

Thankfully, there are nerds all over the planet who do this kind of painstaking, detail-driven stuff for pleasure and take a great deal of satisfaction in checking each other's facts, making each other curious and trying to catch each other out. It's what makes science go. They studied the evidence embedded in the ice and the rocks, the seabeds and the coral, like forensic teams studying the longest, slowest crime ever committed. Word came in from a wide spread of scientific disciplines, from places all over the world; and the evidence was damning.

The warming was being caused by a rise in carbon dioxide. And the rise in carbon dioxide, and other gases related to the burning of fossil fuels, was being caused by us. We were burning our way towards a climate catastrophe.

Friends of the Earth Ireland

Activists all over the world are joining forces to ask the United Nations to pass a treaty on fossil fuels so countries will promise to stop extracting fossil fuels, phase out existing stockpiles, and fast-track clean energy solutions like renewables. Lots of people are joining this action – calling on their cities, regions, and countries to join this treaty. Have a browse at Campaign.FossilFuelTreaty.org

CHAPTER 7
THE SKIN OF THE APPLE

Humans have lived through massive climate change before. The people who existed ten thousand years ago, during the last ice age, were physically the same as us. Even at this primitive stage, their ability to adapt to different environments had allowed them to survive in a wider range of habitats than any other animal. They migrated across the continents, using tools and fire to provide themselves with food, clothes and shelter. With a drop of five degrees and more, our climate changed dramatically, and that change transformed the planet. And when glaciers stretched across the Earth, humans found ways to survive in areas where many other species died away. Despite all our intelligence, imagination and capacity for exploration, however, we have now reached the absolute limits of our planet. We've even started to move beyond them.

From where we are today, a *rise* of four or five degrees, which could be possible by the year 2100, would be a total transformation in the other direction. The world would be as unrecognisable to those of us living in this era as today's climate would be to people living during the ice age.

Even with all we know, it still seems extraordinary that human beings could have the effect we've had on our planet. Up until now, I've mostly been talking about the changes in the atmosphere.

Though the atmosphere wrapped around our world is hundreds of kilometres thick, it's held in place by gravity, which pulls most of it close, so the gases are concentrated near the bottom. Most of the air in our atmosphere is within the first fifteen to twenty kilometres. After that, it starts getting pretty thin. Humans can't survive without oxygen tanks anywhere above eight kilometres.

These days, we can go places our ancestors could only dream of, and yet our world is still very limited. If you started at an altitude of eight kilometres and dropped down to the surface of the ocean, and you kept descending to the deepest point, a place called the Mariana Trench in the Pacific Ocean, eleven kilometres down, you'd have travelled through the full depth of the world in which life exists – less than twenty kilometres. And yet, most of *that* can only be explored by humans if they're travelling in specially designed vehicles.

So you see, the environment within which we can survive without artificial support is a thin, delicate coating around a giant ball of lifeless rock floating in the vacuum of space. If the Earth was an apple, all the life that ever existed on it would have lived within that single layer of skin around the outside.

That's all we have. It's all the environment we'll ever have on Earth.

There's something out there in the sea

The oceans hold about 1.3 billion cubic kilometres of water, which is a *lot* of water. The largest types of ships in the world are oil tankers – y'know, because of all the oil we use. They are mere specks that can be lost in that vast expanse. That water, with its massive, heaving currents, heats up much, much more slowly than the air. And a dangerous secret lies hidden in those ocean depths. It doesn't show on the surface, but the sea has been absorbing most of the extra heat we've been hit with over the last century. In fact, the National Academy of Sciences in the United States has calculated that the Earth's oceans have absorbed more than 90 per cent of the heat gained by the planet between 1971 and 2010. And, incredibly, *the sea is warming up*. I mean, the whole sea ... like, all of it.

Trying to imagine how much heat it would take to warm 1.3 billion cubic kilometres of water by even one degree would probably just make you cry, so let's put it in some kind of perspective. From the National Academy of Science's figures, the amount of *extra* energy absorbed by the oceans is equivalent to that generated by roughly one and a half Hiroshima-size atomic bombs exploding *per second* over the past 150 years. This is what

happens when the whole planet holds onto just a little bit too much of the sun's power.

It is also, however, a great illustration of the power available to us if we could harness that solar energy, and why it is the fastest growing energy industry in the world.

When it comes visiting

These kinds of figures are so huge, you can't really get your head around them. And what does it all mean? How do these changes affect our daily lives? Where can you actually see anything happening? One consequence of warming oceans is that sea-levels rise, and the kinds of surges caused by storms get worse. The sea rushes in against the land, higher and with greater force. In many such places, the threat has been acknowledged for a long time, and long-term projects are already taking action to meet that threat. An example of this can be seen in London.

Which you might think odd, as London is not even on the coast.

However, its main river, the Thames, is a tidal river; the sea flows upstream at high tides and that causes the river level to rise, which can be made even worse if heavy rainfall means that there's more water coming downriver too. Both of these events are completely natural occurrences, and happen in many other places around Britain and Ireland too. But it's a particular problem in London because it's such a huge

city, and large parts of it are on low-lying land. If the Thames were to burst its banks, which was a much more common event in the past, those areas of London could be badly flooded.

Today, major efforts are taken to control the river levels, and the most important, and most visible, is the Thames Barrier. Built in 1982, it is a seriously impressive piece of engineering that stretches from one bank to the other, and its ten enormous gates can be lifted to create a wall across the river. The barrier is closed when there is a risk of flooding, and the number of closures varies every year. Looking at it from one year to the next, there doesn't appear to be much of a pattern, some years spike and others are uneventful. And yet when you study it over the full life of the barrier, there's a detectable rise in how often it gets closed.

Britain's Environment Agency uses the number of Thames Barrier closures as a way of assessing the risk of floods in London. Its records show a steady increase over the decades: just four closures in the 1980s, thirty-five in the 1990s, and over a hundred since 2000. It has been extremely effective in controlling the level of the river. The people in charge of the barrier, however, are already planning for the future, and they say London is going to need a new, bigger barrier to cope with what's coming in the years ahead.

Fortunately, this technology has continued developing since the Thames Barrier was built, as the world of engineering takes on the challenge of adapting our cities to rising sea levels. In the Netherlands, where large parts of the country lie *below* sea level, they have some very sophisticated systems for holding back the sea. One example is the Maeslantkering, the world's largest storm surge barrier, which protects the port of Rotterdam and the surrounding area. Its giant gates can block off a waterway three hundred and sixty metres wide – the equivalent of laying the Eiffel Tower on its side.

For decades, all over the world, governments have been stepping up to meet the challenges that climate change has already started throwing at them.

Across the Atlantic in the United States, the state of Florida is also facing into the reality of rising sea levels. The home of Disney World, Universal Studios, The Wizarding World of Harry Potter and SeaWorld – as well as about twenty million people – has found itself right on the front line. Their problems are made worse by the fact that much of the rock on which their coastal towns and cities are built is a porous limestone and sea water can seep through it – right under walls and buildings. It also means that sea water can contaminate their water supplies. The state is fighting back, planning over four billion dollars in sea-level-rise solutions, which include protecting sewage systems, building up roads, improving storm-water drainage, and erecting sea walls.

Bangkok, in Thailand, is another low-lying city that's already suffering regular flooding, and is likely to be partially submerged within the next fifty years. But they're not hanging around waiting for it to happen. The measures they're taking include cleaning out and widening their network of 2600 kilometres of canals to absorb the impact of floods, along with pumping stations and drainage tunnels.

All over the world, countries are putting massive amounts of money and resources into holding back the sea or adapting to increased flooding. It's a demonstration of how seriously they're taking the threat of rising sea levels. Like the rise of a few degrees in temperature, it only takes a slight rise in the levels in our oceans to have some really, really big effects. And with so many of the world's major cities on or near the coast, those effects will have all the more impact.

Imagine a bath that's full to the brim. As long as no more water is added, it won't overflow. But once you start adding water, even a little bit, it doesn't matter how deep the bath is; it will start to spill out over the edges. Our cities are right up on the edges of the water. Water expands as it warms. Though the oceans are vast, they are brimming up over the edge, and melting glaciers dropping trillions of tonnes of ice into the sea aren't cooling it, they're just adding more water. Slowly, ever so slowly, the water is starting to spill over. Thousands of years ago, when humans faced major environmental changes, they just picked up and moved. And those changes happened very slowly.

This time, the changes are coming at us much faster, and you can't just pick up a city and move it. On a geological time scale, our cities are sand castles on a beach. Something's gotta give. This is a challenge we have to meet, not just with great feats of engineering, but with the planning for our towns and cities, the operation of our ports and rivers, and even landscaping sections of our coastlines to absorb and withstand the impact of storm surges.

And it's not all about us either. There are changes happening out in the oceans that are affecting far more than just the water.

Sea-swimming and spending time in nature is known to have positive effects on our physical health and mental wellbeing. When we connect with nature, we're reminded of how good it is for us. The Earth gives us everything we need, and we should be its caretakers.

CHAPTER 8
BENEATH THE SURFACE

The life that first created Earth's oxygen all those millions of years ago began in the sea. And for all the talk of the rainforests and how they're the lungs of the world, most of our oxygen still comes from the oceans. And most of that comes from phytoplankton, single-cell plants that use photosynthesis just as plants on land do. And, of course, they also absorb carbon dioxide. A good chunk of the carbon they take in will eventually sink to the ocean floor when the phytoplankton die, which means it doesn't end up back in the atmosphere.

Unless someone happens to drill down into it a million years later, when it's turned into oil.

Like plants on land, phytoplankton are the base for life in the oceans. They are the foundation that every other type of creature – from zooplankton to blue whales, from bottom feeders to peak predators – relies on to survive. If these single-celled plants died off, the creatures that eat them would soon follow, and the creatures that eat those creatures would be next – all the way up the chain.

The ecology that they support would collapse, and we would lose most of the life in our seas. And more than half the Earth's oxygen supply. Oh, and they'd stop absorbing carbon dioxide too.

For most of our existence, humans have understood very little about this ecosystem beneath the waves; our exploration of this awe-inspiring underwater world was limited to how long we could hold our breath. While the first diving bells were developed centuries ago, practical diving suits with hoses weren't in use until the 19th century, and then it wasn't until the 20th century that divers could start carrying their own air tanks or driving submarines into the depths.

Scientists have been making up for lost time; our knowledge of the oceans has expanded enormously since those early days, and more discoveries are made every year. Still, there's so much we haven't seen. It's a bizarre fact, but most of the floor of the sea is still unmapped. We know more about the surface of Mars than we do about the ocean floor. There is a planet-wide landscape of mystery down there, waiting to be discovered. And the more we learn about this place, the more we understand how life on land is dependent on those ocean depths, and the more we can do to preserve this ecosystem.

Getting the message

The life in our seas is one of the best ways of detecting change in this vast, vast expanse of water that's wrapped around our planet.

In the early 19th century, workers in coal mines used birds such as canaries in cages to detect carbon monoxide and other toxic gases. These are known as 'sentinel species'. As they were more sensitive to these gases than humans, they would be affected first, serving as a warning to the miners. This practice continued right up until the 1990s. The birds were more than just a warning system; they were companions too, treated fondly as pets by men who faced gruelling work in dangerous conditions. Life in our seas is the canary in the coal mine that is Earth's environment. We need to pay as much attention to our sea life as those miners paid to their birds.

The life in the oceans can be very sensitive to changing conditions, and that life is not distributed evenly. There are places that positively fizz with creatures, and other areas where little or nothing survives. The most diverse habitats in the oceans are the coral reefs. Built by tiny organisms known as coral polyps, these reefs are the rainforests of the seas, a gorgeously rich, complex and colourful mass of life, existing in a cycle that ensures there's food for everyone, including the humans who fish there.

The most famous of all is Australia's Great Barrier Reef. Stretching over 344,400 square kilometres, the reef is located in the Coral Sea, off the coast of Queensland. A metropolis of nature, it draws tourists from around the world, who come to see the extraordinary variety of creatures to be found there, including turtles, whales, clownfish, stingrays, octopuses, sea snakes, sea anemones, giant clams, sharks and thousands of other species all doing their thing. It's like a Pixar film, but in real life.

All this life stems from the relationship between the coral and the algae they live with. The algae provide food for the coral, which in turn provide shelter and nutrients for the algae. Small creatures also come to feed on these, which in turn attract larger fish and so on until you get to the great white sharks and the most dangerous of the lot, human beings. It is a thriving network of food.

Coral can be killed off by pollution, storm damage and predators like the crown-of-thorns starfish. They are also vulnerable to changes in temperature. Extended warm periods can kill off the algae the coral polyps feed on, leaving them to starve, a process known as 'bleaching', because the death of the colourful algae leaves the sticks of coral looking like bleached bones.

You might think they must be very delicate that such a small change can kill them off, but bear in mind that if your core body temperature went up or down by three degrees Celsius, it would kill you too. It's basic biology; we are all vulnerable to changes in temperature.

This kind of mass die-off starts a cascade of starvation, as one after another, the different levels of creatures that feed on the reef run out of food. Since 2016, *half* of the coral in the Great Barrier Reef has been killed off by bleaching caused by warming seas, and this is expected to get worse as time goes on. For marine biologists and other people who study the sea, this is the equivalent of a deafening fire alarm. Life in the oceans is in trouble.

And it's not like we've been helping; a huge international ten-year study of the world's seas entitled *The Future of Marine Animal Populations*, carried out between 2000 and 2010, found that fishing had already wiped out 90 per cent of large predators. They also found that, compared to 1950, the ocean had 40 per cent less phytoplankton, and that human activity was damaging coral reefs as well as increasing the risk of marine populations going extinct. Quite separately from the whole climate change thing, we've also been slowly destroying one of our major sources of food ...

And, y'know ... *oxygen*.

If there is one positive to be drawn from this, it's that the Great Barrier Reef is so iconic, and so many people have an emotional attachment to it, that it has become a very visible symbol of how our world is changing. People love this place, it is precious to them and they want to save it. We cannot visualise a changing world, but we can see the fading beauty of one of the planet's most famous wild environments and it does more than any explanation of the science to motivate people to act.

Science is already providing one solution. Placing metal frames on the seafloor and running a weak electrical current through them causes a thin coating of limestone to form on the metal, which is a perfect foundation for coral to build upon. This technique can be used to repair existing reefs or even start new ones. The polyp has been found to grow three to four times faster than normal on these artificial reefs. The electricity can be provided by solar panels on rafts, the structures grow stronger with age and are self-healing. The technology is so simple, it's being reproduced by small coastal and fishing communities, customised to local needs, as well as on areas of the Great Barrier Reef. It's still experimental, but is being used with increasing success. People are using electrified metal to rebuild ecosystems – which is pretty amazing.

Bad chemistry

With all the changes that are happening to our seas, one of the things that doesn't get a lot of attention is that they are becoming more acidic. Don't be alarmed – this doesn't mean the water's going to start eating away at your legs when you walk into the waves on your next visit to the beach. The oceans absorb nearly a third of the carbon dioxide in the atmosphere, and it causes a chemical reaction that lowers the pH of sea water. The pH scale is how we measure the strength of acidity. It goes from 0 to 14. A pH of 7 is neutral. A pH of 0 is really acidic; an example at this end of the scale would be battery acid. A pH of 14 means it's a really strong base – which is also really, really corrosive. Drain cleaner is an example of a strong base.

Over the last two hundred years or so, the seas have become about 30 per cent more acidic, although because they weren't very acidic to begin with, a pH of about 8.2, you wouldn't notice any difference if you went swimming in it. It's not like it's going to dissolve your swimming togs or anything.

However, if you live your whole life under water, it can be a whole other deal. The increase in acidity can cause problems for any sea creatures that need to grow shells, like sea urchins, clams, oysters and coral, along with plenty of others, including many of the micro-organisms that, like phytoplankton, make up the base of the food chain. When creatures like these fail to develop properly, they affect everything that eats them too.

Acidification can also affect the development of the young of larger fish, and can interfere with their senses, making it more difficult for them to tell the difference between their own species and predators, or even the foods they can eat. These tiny changes can have a major effect when they're multiplied across huge fish populations. Like the rise in carbon dioxide in the atmosphere, it's the kind of change that would have taken thousands of years in the past. This time it's happened in less than a couple of centuries. Scientists still don't know what this means for the future of the oceans, but there is a lot of research going on in this area to find out.

Dead zones

Another phenomenon we're seeing in the oceans is what are known as 'dead zones'. These are areas of water that don't have enough oxygen to support life. They can and do occur naturally, though many are now being caused by humans. When rain washes fertiliser from our fields into rivers, it can be carried down to the sea to where it feeds algae in the sea water. This causes 'algae blooms', explosions of short-term, massive growth, as if the stuff is on steroids. Then when the algae dies, it sinks to the bottom and decomposes. As it rots, it soaks up all the oxygen in the water. Off the coast of Texas and Louisiana, at the mouth of the Mississippi, the river's mighty flow has emptied fertilisers and sewage into the sea, creating a dead zone in the Gulf of Mexico with an area of more than 20,000 square kilometres.

It's worth noting that, compared to changes in our climate, this is a problem that can be tackled relatively quickly by reducing the amount of fertilisers and waste we're dumping into our rivers.

While climate change is not causing dead zones directly, it can make them worse. Warm water can cause the blooms to happen earlier in the year. It also holds less oxygen and is less likely to mix with the colder, denser water below, so the problem lasts longer.

And while this is an important element in the life of our oceans, there's a much, much bigger movement of water that the life of our planet relies on.

The great ocean conveyor belt

We saw earlier how the sun's heat and the turning of the Earth create the air currents that distribute that heat, and it's this process that gives us our weather. However, there's even more to it than that, because the oceans also play a major, but much less visible, part in our climate. The water in the oceans is constantly moving, and like the blood pumping around our bodies, this movement is vital to the health of our planet.

The winds affect shallow currents, down to a depth of about a hundred metres or so, and these currents tend to be quite fast. One of the strongest currents carries warm water from the equator up towards the North Atlantic. As it travels up there, it loses heat, until it gets so far north that sea ice starts to form. The freezing water loses its salt, making the water around it saltier, which also makes it denser and heavier. This cold, salty water starts to sink, drawing in the warmer water that's coming up behind it, keeping the cycle going.

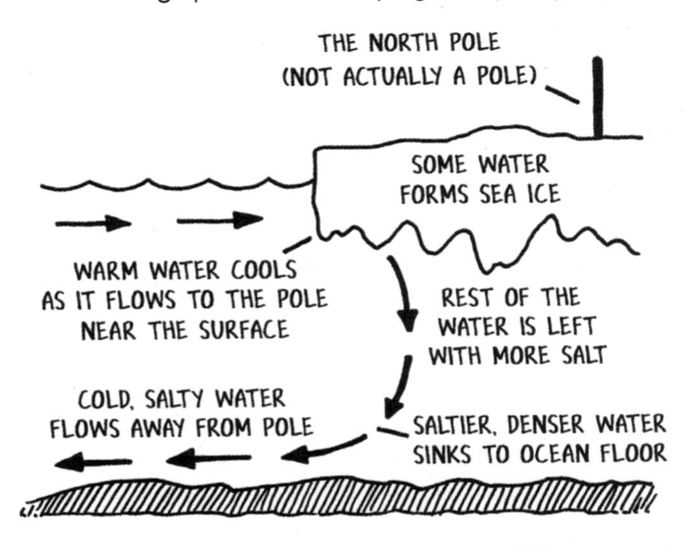

That cold water now starts a very different and much, much slower journey, heading south along the Atlantic again. Like the world's worst rollercoaster, it crawls down around Antarctica and up to the equator in a dense mass, thousands of metres below the surface. It will keep going until it has lost much of its salt and gained enough heat to rise again, some of it in the Southern Ocean and the rest in the Pacific or Indian Oceans. And when I say this current is *slow*, I mean it. The 'oldest' water will have taken up to a thousand years to complete one lap of this course. From there, the whole process starts again.

This weaving loop of currents is affectionately known as 'the great ocean conveyor belt', and it's how the seas spread heat around the globe. The technical name for that deep, slow current is the *thermohaline circulation* – 'thermo' for temperature, and 'haline' for salinity, or saltiness, because it's the changes in heat and salt that keep that powerful flow of water pumping around the Earth.

It was mentioned earlier that the entire ocean was warming up, but this is not *quite* true. There are a few areas that are stubborn exceptions to this. One of them, in the North Atlantic, is what's known as 'the cold blob'. (Seriously, you have to love these technical terms.) It's an area south of Greenland, and it has dropped in temperature by nearly a degree Celsius, sometimes referred to as a 'hole' in the warming. Research suggests that this is being caused, at least partly, by a slowing of the conveyor belt at this point, affected by the fresh water draining off Greenland's glaciers.

For those of us living in Ireland and Britain, this conveyor belt of heat has a hugely important effect on our lives. The part of this current that passes our coasts is called the North Atlantic Drift, and along with our prevailing winds, it helps warm up our region. Without it, our climate could be more like Canada's, and winters would mean months of snow, frozen rivers and ice in the seas around our coasts. It's just one of the many ways that what we think of as normal life relies on the flow of the oceans.

It's worth remembering that there's only a few degrees in the difference between our climate and North America's. There was a time only ten or fifteen thousand years ago – a blink in the eye of Earth's history – and five or six degrees lower in global temperature, when *both* North America and Europe lay beneath the crushing weight of glaciers. Some glaciers from that era still exist in the world today, and they have a lot to tell us about what's happening to our planet.

Friends of the Earth Ireland

Joining a beach-clean or a park litter-pick is a great way to get out in nature while tackling waste. Record how much waste you collect and share the data with your county council to encourage them to do more on waste in your community.

CHAPTER 9
THE ICE GIANTS

Glaciers are *old* and *slow*. It takes a serious amount of time to make one. To get a sense of how long it takes to form a glacier, you have to think about snow falling on the ground, because that's where glaciers come from. Snowfall. Which doesn't seem right because some of them are *kilometres* deep. Normally starting on high ground, such as mountain ranges, snow settles in layers, one day after another. Over time, it crunches down under the weight of the snow that follows and becomes more compact, until it's more ice than snow, and one layer builds into the next. That's all it takes, regular snowfall – building up over several thousand years.

Glaciers can be found all over the world, but most glacial ice is found in Antarctica and Greenland. Some can be the size of football pitches, while others can be hundreds of kilometres long. The ones we have now date back to the last ice age and beyond. And because they've formed on high ground, and they're kind of heavy, and they're made of ice, they don't stay put. Gradually, gravity takes effect, and these giant masses of ice start to slide downhill, like a river in slow motion. In warmer regions, they melt when they reach lower levels and actually turn into rivers. In some parts of the world, they are the main source of water for the people who live there. Other glaciers reach the coast and extend out over the sea, forming ice shelves. Their leading edges regularly break into pieces and float away as icebergs.

Glaciers are huge, so they move very, very slowly – maybe only a metre or less a day – cracking and splitting and re-forming as they follow the landscape, grinding over the stone, making their way through the mountains. And as the lower end is trying to head downhill, the higher surfaces are still collecting snow. The size of a glacier increases and decreases based on how much snow falls and how fast its leading edges melt, but there is clear evidence that the world's glaciers have been steadily shrinking since the Industrial Revolution.

Stories trapped in ice

One of the wonders of this old ice is that, like earth and rock, its layers can be read by people who speak its language. Scientists can drill down into this ice and pull out long columns, known as ice cores. The ice near the bottom of some glaciers is hundreds of thousands of years old, and the different levels of ice along the cores can be analysed, particularly the tiny bubbles of air trapped there. Scientists can study the very molecules in this air and learn all sorts of things about how the planet's atmosphere has changed through the ages – including things like global temperatures and carbon dioxide

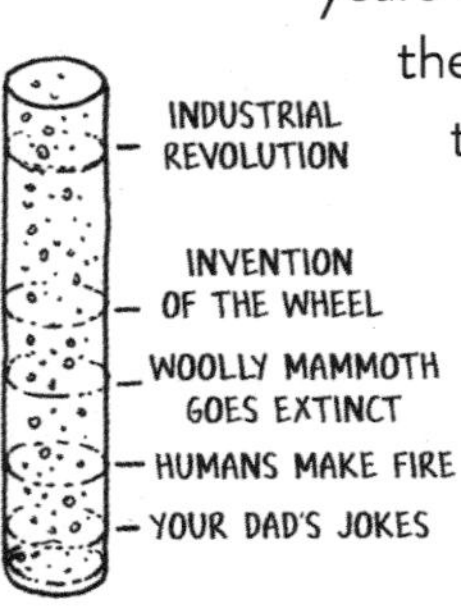

levels. Imagine being able to study air that was breathed by creatures 100,000 years ago. Like the compacted layers in rock and in the seabed, this ice holds ancient memories of the Earth.

It is information like this that climate scientists use to piece together the story of our world, and how we are affecting it. And yet, it is not just in these tiny scientific details that we can find evidence of our changing climate, because the glaciers have a much more dramatic way of getting our attention.

The bigger they are, the harder they fall

Glaciers melt, and they break apart and they drain into the sea; it's a natural cycle. Now, however, they're doing it faster than ever before. On the 1st of August 2019, the Greenland ice sheet broke its own record for the amount of ice lost in a single day. It sent an estimated 12.5 billion tonnes of ice pouring into the ocean. That's just *one day's worth* of ice, and just from Greenland's glaciers. Head-wrecking facts like this give us some idea of why the threat of rising sea levels is a very real problem.

Something the size of a glacier doesn't die quietly – or quickly. If a glacier experiences a few days of sudden melting high in the mountains, water can rush down in rivers and streams, flooding the land below. It can create an even bigger disaster if it creeps across a river, blocking it and forming a lake. When this dam of ice weakens and collapses – as it eventually will – it sends a deluge of water, ice and debris downriver. If a ridge of ice is jutting over a steep mountainside and becomes unstable, pieces can break off and cause avalanches.

And then there are the icebergs. While it might sound like a contradiction, global warming is leading to more icebergs. The more ice that melts off places like Greenland and Antarctica, the more icebergs will end up in the sea, and they'll be bigger too. We often remember the *Titanic* when we think of icebergs – in fact, I had nightmares about icebergs when I was a kid after watching an old film about the disaster. Looking back now, icebergs weren't as big a problem in my life as I thought they might be.

And times have changed. These days the big icebergs are closely monitored by satellite, and ships are steered well clear of them. Even so, they can be extremely dangerous and cause major problems to shipping. And some of them are colossal. The largest iceberg ever recorded was named the B-15. It broke off the Ross Ice Shelf in Antarctica in the year 2000, and was 295 kilometres long and 37 kilometres wide, with a surface area of 11,000 square kilometres. It was larger than the island of Jamaica.

Try and imagine the problems caused to wildlife, shipping and coastal communities by having a slab of ice the size of a country floating across the ocean. While this was an exceptionally large berg, it does represent the scale of ice that will continue breaking off our glaciers. And because of their size, these mountains of ice can hang around for a long time. Large pieces of B-15 were still being tracked eight years later.

The deflector shield

It's not just the glaciers that are melting; the sea ice is too. While glaciers form on land, this is ice that has formed at sea and floats on its surface. Unlike glaciers, melting sea ice does *not* contribute to rising sea levels, as it's formed from sea water and stays floating in the sea. The largest and most important mass of sea ice is in the Arctic Ocean, centring around the North Pole, and it's constantly moving, melting and re-forming over the seasons. There's no land up there, it's just ice; if you went on an expedition to the North Pole and planted a flag there to mark your achievement, the flag would have moved by the following year. You can imagine the disappointment.

The Arctic is warming twice as fast as any other part of the planet, and this is affecting the formation of new sea ice. Less of it returns each winter and it's possible that, during the summers, we'll see the complete loss of Arctic sea ice in this century. The disappearance of the ice cover is already affecting the ecology of this region, threatening the food chain for creatures like polar bears, wolves, seals, whales, many types of fish and a range of others.

What is even more important for the world as a whole is that the vast expanse of white ice over that part of the ocean acts as a heat shield, reflecting sunlight back out to space, helping cool that area of the planet. This is called the 'albedo effect'. Bright surfaces *reflect* light, dark surfaces *absorb* it. As the darker ocean shows through the ice, it absorbs more light, more heat. The Arctic starts warming up, which makes the ice melt faster, which leaves less ice to reflect the sun's radiation, which causes the warming to increase, and ... well, you get the idea.

Remember how the creation of ice is one of the things that powers the thermohaline circulation? Water freezing into ice loses its salt, which makes the water around it saltier, denser, so that water sinks, driving that long slow flow that crawls along the depths of the oceans. If less ice forms in the Arctic Ocean, it could slow down the great ocean conveyor belt, that looping network of currents that distribute heat and life around the planet.

And on top of all this, that extra heat building around the Arctic Ocean is already having serious effects on our weather.

Friends of the Earth Ireland

While exploring the Antarctic in the 1950s, French glaciologist Claude Lorius discovered how gases are trapped in the ice when he saw bubbles coming from the 1000-year-old piece of ice core in his whiskey. He realised the connection between these gases and climate change and became an advocate for climate action. Watch the documentary Ice and the Sky *for more.*

CHAPTER 10
TURBO-CHARGING THE WEATHER

There have always been extreme weather events and humans have experienced some humdingers over the course of our civilisation. It's extremely difficult to trace a single event back to our changing climate, though there is a whole new field of 'attribution science' that focuses on this very task, to advance our understanding of it. What scientists can say for certain is that global warming increases the *risk* of extreme weather events, and there has been a well-documented increase in the strength and frequency of such events across the world.

It might seem strange that global *warming* can cause *cold* weather, but you have to remember that heat is also energy – and that pumping more energy into our atmosphere and oceans gives them the power to do more, enabling our weather to swing out further to the extremes. Charged up on their version of an energy drink, our weather systems go into overdrive, regions of high and low air pressure shifting around faster. The bigger differences between these areas of pressure make winds stronger and weather more changeable; hotter air means more water evaporates from the oceans, which influences the strength of storms. It is a machine of mind-boggling complexity.

How does heat make cold weather?

This can happen in different ways, but one example of how rising temperatures can lead to cold weather was demonstrated by 'the Beast from the East', a snap of severe cold that hit Ireland and Britain in February and March 2018. It was our coldest weather in decades. Snow lay thick on the ground. There was much surprise and alarm and, in a mad rush of panic buying, all the shops ran out of bread. It was caused by a thing called a polar vortex, a low-pressure mass of freezing cold air that rotates over the North Pole, spinning much like you'd see a hurricane doing on a satellite image. This cold air is normally kept contained up in the Arctic by the northern polar jet stream, a powerful air current that flows west to east around the top of the globe at hundreds of kilometres an hour. You'll probably have heard of the jet stream because it's that high-altitude path that jet airliners ride on long journeys. It's like a river, fast-flowing, powerful and reliable. It gets that power largely from the difference between the cold at the poles and the heat at the equator.

However, in 2018, the North Pole suffered what's known as a 'sudden stratospheric warming event'. The air high in the atmosphere above the pole warmed up – it was still *freezing*, but it was warmer than it should have been at that time of year. This weakened the jet stream, causing it to wobble, and that freezing cold air it was fencing in started to wander. It headed southwards. Instead of the mild dreariness we'd expect at the end of winter, we were hit with weather from the Arctic Circle. So basically, we got freezing weather because the North Pole warmed up a bit and started leaking.

This happened across North America too, where they faced even worse Arctic conditions, including record-breaking blizzards, also caused by the polar vortex. It was worse there than in Europe because weather extremes change depending on where you are on the planet.

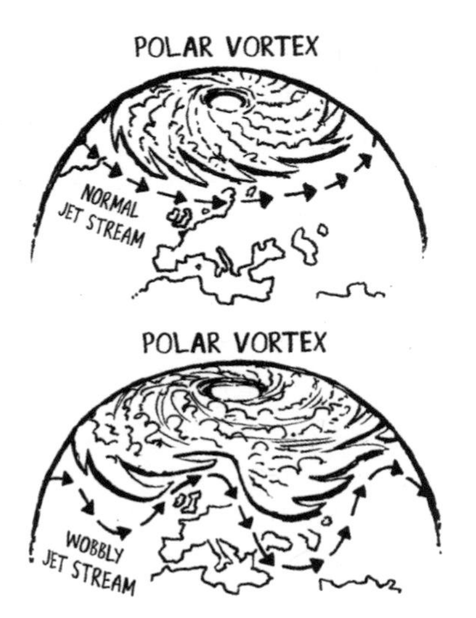

Storm warning

A hotter atmosphere combined with warmer oceans means that more water evaporates from the sea. Heat and moisture are the fuel for storms, including hurricanes or typhoons. It's why they start out over the sea. This can result in powerful, damaging winds, heavy rainfall and, on the coasts, flooding from storm surges. On higher ground, it can lead to flooding as rain causes streams, rivers and lakes to overflow. Hurricanes sweep across the Atlantic and the Caribbean, wreaking havoc over the islands and the eastern coasts of North and South America, particularly places like New Orleans, Florida, the Bahamas, Mexico and Puerto Rico.

In the Pacific and Indian Oceans, where the same storms are known as typhoons or cyclones, it's countries like the Philippines, Bangladesh, Myanmar, Laos, Vietnam, India and Japan that take the biggest hits. Storms on this scale don't just harm people and damage property; they can cause disasters that can take years for a country to recover from.

The physical geography of any area has a big influence on the kind of weather it experiences. Of all the places in the world, the US central plains suffer the most tornadoes, but these dramatic twisters can turn up in other regions too, including Europe, Australia, Southern Africa and parts of Asia. The Daulatpur–Saturia tornado in Bangladesh in 1989 is thought to have killed as many as 1300 people, making it the deadliest tornado in history. Capable of creating winds of nearly five hundred kilometres an hour, these short-lived phenomena can dig trenches through the ground, turn loose debris into deadly missiles and pull buildings from their foundations.

While winds can cause all kinds of destruction, the greatest storm damage is caused by flooding. In some cases it comes in the form of surges from the sea; in other cases, it's purely down to rainfall. In July and August of 2010, record levels of monsoon rains in the northwest region of Pakistan resulted in flash floods that swept away buildings, roads and bridges and drowned farmland. The floods affected about twenty million people. With huge areas of crops destroyed, millions went hungry.

When fields are flooded, it can sweep wide swathes of earth into the rivers. This soil erosion not only deprives the farms of valuable fertile earth, it can also cause blockages in the rivers that increase chances of them breaking their banks. It's estimated that between 1200 and 2200 were killed, while over a million houses were damaged or destroyed. It was one of the worst catastrophes in the country's history.

Dramatic events like severe flooding can be bad enough, so it seems unfair that the same countries that suffer some of the worst floods can also be hit by the worst heatwaves.

Too hot to handle

When we think of a world warming, cold weather and storms might seem odd, but it's easier to understand an increase in heatwaves. In Ireland and Britain, we think of heatwaves as a reason to head for the beach, eat ice cream and bake in unhealthy levels of solar radiation, but even in places that enjoy moderate climates, heat can cause major problems. In June of 2019, a heatwave struck across Europe, just the latest one of many in recent years that are serving as a clear sign of our warming climate. France experienced a new high-temperature record of 46°C, and records were also set in the same month in the Czech Republic, Slovakia, Austria, Andorra, Luxembourg, Poland and Germany.

The highest ever recorded temperature in Europe was 48°C (118.4°F), measured in Athens in July 1977. On average, however, the twenty warmest years on record have all been within the past twenty-two years, with 2015–2018 making up the top four. Heatwaves are dangerous because if a person's body temperature rises above 40°C (104°F), heat stroke can set in. This is where the body overheats, causing damage to vital organs that can eventually lead to death. Those who are particularly at risk are the elderly, the very young, sick or overweight people or people with other health problems. It is not just a case of 'getting too much sun'. After the severe heatwave in 2003, the rate of additional deaths recorded in Europe went up by 70,000 compared with previous years.

Mainland Europe's 46°–48°C might feel mild compared to what's possible in places closer to the equator, like India and Pakistan, which had a catastrophic heatwave in 2015 when the monsoon rains were late to arrive. It got so hot in Delhi, the surfaces of the streets began to melt. Sections actually turned to liquid, with road markings oozing into new shapes. There was such demand for electricity to power fans, water pumps and air conditioners that power failures became common, which of course meant that none of these things worked when they were needed most.

Countries around the Persian Gulf, such as Qatar, the United Arab Emirates, Saudi Arabia and Bahrain, where humidity (moisture in the air) can make a hot day feel even hotter, could regularly be facing temperatures over 40°C; but with humidity of up to 100 per cent, it will *feel* more like 70°–75°C. In conditions like that, the body can't even cool itself by having sweat evaporate, so even a fit and healthy person will quickly begin to overheat. It won't be safe to go outside during the day, and having air conditioning in your home could become a matter of life and death. And running air conditioning all day and night is expensive. There are many who can't afford it.

One solution to this is to open public spaces that offer relief when things get too hot. In the United States, some cities now operate 'cooling centres' in public buildings like community centres and libraries, as well as parks and recreation sites, where people can escape to on the hottest days to chill and stay hydrated. Apart from the physical relief they provide, these centres can contribute to a community spirit in difficult times. In fact, in countries all over the world, it is the combination of communities and their public services working together that's providing much-needed hope and solutions for people trying to adapt to their changing environment.

As with extreme conditions in every area of the world, the poor will be hit hardest, but *everyone* will be affected. And when it comes to dealing with hot conditions, access to water is of the utmost importance.

Going dry

It probably comes as no surprise that an increase in global temperatures will cause shortages of water. Heat and deserts go together so naturally in our minds. However, any region can run short of water, especially when you're used to having a certain amount in your streams, lakes and rivers and then they start running dry. This is true even in countries like Ireland or Britain, where we're used to having more than we need – or even more than we'd like.

It's when you start running out of it that you realise just how vital it is, and in some countries, that's more common than others. Just like blizzards, floods and tornadoes, drought is affected by geography. And the first impact can be on farming, so not only is your water supply affected, your food supply is threatened too.

The *Guinness Book of Records* lists the Chinese Drought Famine of 1876–79 as the worst in human history. The rains failed for three years in a row, and between nine and thirteen million people died. People tried to eat earth and the leaves of trees, and dead bodies could be found lying along the roads. At around the same time in India, more than five million people starved when the monsoon failed.

Any major famine these days will now trigger an international response, with countries from all over the world stepping in to offer supplies and aid, as well as agencies like the UN's World Food Programme and the International

Federation of Red Cross and Red Crescent Societies. And while we might believe that we can now prevent death tolls like that of the Chinese drought, disasters like this are still happening, we seem unable to prevent them, and they can often be made worse by foreign interference and war. And their effects, wherever they strike, can be devastating and long-lasting on the populations who endure them.

This brings me to a point I want to make about the power of stories. While we have the means to communicate around the globe in seconds, there is a major difference between knowing something and caring enough to take action. We hear about crises and disasters in other parts of the world all the time. Some have more of an emotional effect than others. Some provoke more empathy than others. The issue of climate change suffers from this problem because, all too often, it is hard to relate the big picture facts to the day-to-day life of a person we can empathise with. We need to hear someone's story, to make it personal, to get us emotionally involved.

The worst crisis in Irish history was the Great Famine of the 1840s. The poor in rural Ireland survived almost entirely on the potato; it was the most nutritious crop you could grow for the area of land it required. When a disease called potato blight struck in the warm, damp conditions of 1845, the potato crop rotted in the ground all across the country. The Irish began to starve, a situation made worse by the mismanagement by the British government. Across the Atlantic, in the United States, news of the famine reached the people of the Choctaw Nation. They were in a desperate situation themselves. The US government had driven them from their homeland in Mississippi to modern-day Oklahoma. Thousands had died on what became known as 'The Trail of Tears'. Though they had very little to give, they held a collection and sent $170 – about $5000 in today's money – to help the Irish.

Over 170 years later, this is still a well-known story in Ireland. There is a sculpture of steel feathers in Midleton, in Cork, called *Kindred Spirits*, to commemorate this act of generosity. When the coronavirus started to spread across the United States in 2020, the people of the Navajo and Hopi tribes of Arizona, Utah and New Mexico suffered some of the worst effects, because so many lived in poverty and had little access to clean water. A campaign began to raise money for these communities – and the Irish remembered the kindness of the Choctaw Nation. Donations began to flow in, not from big aid agencies, but from individuals. About 26,500 Irish citizens are reported to have donated more than a million dollars towards the campaign, which raised about six million dollars in total, and was used to distribute clean drinking water and food to those in need. Many of the Irish people mentioned the story of the famine when they made their donations.

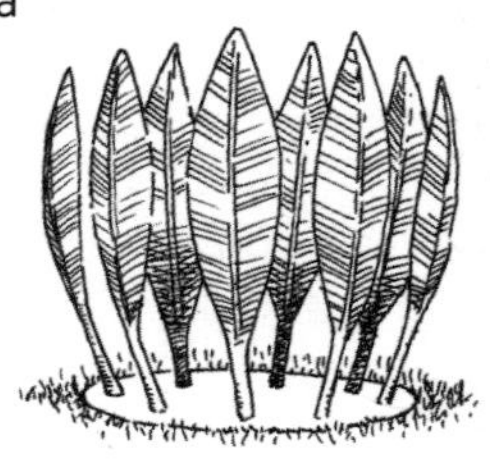

Events like this show us why it's important to share our stories. Stories have power. As is the case with any crisis, we need to find the stories in climate change. It's not enough to know the facts, we have to understand enough to care. We are motivated more by emotion than logic.

Fuel for the fire

Ironically, the same climate conditions that can cause shortages of water can also lead to an increased risk of wildfires – exactly when you have no water to put them out. Fire, our old friend and ally, still has cousins in the wild and we've never quite managed to tame them. When vegetation dries out in drought conditions, it's more likely to catch fire – and that fire will spread faster. The warmer our world gets, the easier it will be to burn.

Towards the end of 2019, it became clear that Australia's seasonal bushfires were getting out of control. Bushfires are a natural phenomenon there, and they occur in many regions, particularly around the south and east of the island continent, in the states of Victoria and New South Wales. Take those long hot summers, dry grass, the oil of eucalyptus trees and then add a dropped cigarette, a careless campfire or a lightning strike and you have enough to start a fire that burns for *weeks*. Add a brisk wind, and that fire can spread faster than a person can run. Bushfires on their own aren't necessarily a major problem. Fire is part of the cycle of life in Australia, the vegetation has evolved to recover from bushfires and the people have learned to adapt to and

cope with them. In fact, the type of fire management used in Australia by indigenous community-based ranger groups, combined with modern technology, is considered one of the best in the world. But in recent years, the fires have been becoming increasingly intense and hard to control.

By January 2020, the news had spread around the world: the country was experiencing the worst fires in its history. By the end of January, dozens of people were dead, and even more died later due to air pollution from the fires. Thousands of homes had been destroyed, over 110,000 square kilometres of land had burned – nearly the size of England – and the smoke was creating a haze over New Zealand, over two thousand kilometres away. More than a billion animals had been killed.

It's not just Australia where wildfires are getting worse. The same thing has been happening in other parts of the world, with some of the worst fires blazing their way through Central African countries such as Angola, the Democratic Republic of the Congo, Gabon and Cameroon. Greece and Portugal have seen unusually bad fires, as have the forests in Brazil. California hits the news every year, and even places not normally associated with hot, dry weather, like Siberia and Alaska, which lie within the Arctic Circle, have been experiencing wildfires. As well as the loss to the environment caused by these disasters, the burning vegetation is releasing millions of tonnes of carbon into the atmosphere.

It's very difficult to link single events to climate change, though this is becoming increasingly possible.

What scientists will do is look at the patterns over years or decades and see how they're changing. And when they see the number and violence of extreme weather events steadily rising, that's when they can start making links with other available evidence.

And just as the changes in the seas have made victims of the creatures living there, so it is with the environment on land. The billion creatures who died in the Australian fires are only one thread in a worldwide web of life that threatens to fall apart if we don't do something to prevent it. And that will take an unprecedented library of knowledge of our world – a library that we are adding to every day.

We know so many things now that we didn't know before. Humans have been studying the weather for thousands of years, and it has been the go-to subject for small talk for as long as we've been able to talk. Around 650 BCE, the Babylonians tried to forecast the weather based on cloud formations. By 300 BCE, Chinese astronomers had divided the year into twenty-four festivals, each associated with a different kind of weather. By the 1600s, both the thermometer and the barometer had been invented, to measure temperature and atmospheric pressure, and other devices were created in the centuries that followed as weather was studied and recorded in more detail. The invention of the telegraph meant that by the mid-1800s different areas of a country could talk about their weather as it was happening. It was now possible to send warnings about weather faster than the wind could carry it, allowing basic weather maps to be drawn up and giving people time to prepare for what was coming.

However, it was only in the 20th century that we really started to analyse and understand what made our weather go, and make long-term forecasts with real accuracy. Starting with ground stations all over the world communicating with each other, we have added weather balloons to carry instruments into the atmosphere, then satellites to look down on it, and then finally created computers capable of modelling weather patterns, playing games with the data, to see what the atmosphere might do under different conditions. This technology has advanced to the point where we are now capable of watching a storm cross the ocean in real time, and we can input the information into a computer that can fast forward that storm and predict where it's going to go.

It is a science that is progressing with increasing speed, and the more we can get ahead of our weather, the more we can take action to reduce the damage it does, use it to our advantage, or even change the weather itself.

Friends of the Earth Ireland

'No change can come about if hope doesn't exist and people are paralysed by inevitability. My best work started when I recovered from my crippling eco-depression and turned it to action, because the sheer amount of people also fighting and creating little pockets of the future they want within the present gave me hope. I am still terrified and furious almost all the time but hope fills the glass that I pour action from.'

Aiyana Hedler, Irish Youth Climate Activist, speaking at the fundraiser for this book

CHAPTER II
SAVING LIVES

When faced with the scale of the problems being created by climate change, we can feel overwhelmed. We are barely able to understand them, let alone do anything to stop them. They can feel distant, untouchable, unstoppable. Some of the changes are locked in: we are so far along the path that they are going to happen no matter what we do now, so we can only adapt to cope with them. But others can still be slowed down and stopped.

The first thing to realise is that nature is not only a means of reading the warning signs, it can also be our most powerful ally. As I'll show in this chapter, we just need to let it do its thing. So much of what humans have done over the course of our civilisation has been in opposition to nature: fighting it, conquering it, taming it and enslaving it. At best, we've treated it as a resource, something to be *used*. At worst, we've viewed it as an enemy to be destroyed. We have needed land to live on, to feed ourselves and create homes for ourselves. As we've hunted, farmed, fished, mined, deforested and built, we have taken up more and more space, claimed more and more land, consumed more and more of that environment, and driven away or killed millions of species of animals. For many years, those who study the plants and animals of our world have been trying to warn us that something is wrong. It's not just that we are pushing species to extinction here and there, but we're collapsing whole ecological systems. As we saw with the coral in the seas, we're knocking links out of nature's food chains – and destroying entire networks of creatures that depend on each other to survive. Biodiversity is thinning out. Micro-organisms, insects, reptiles, birds, mammals and other creatures all have a role to play in the environment that supports human existence, and level by level, our wildlife is disappearing.

Mass extinction

Like blizzards, bushfires, flooding and drought, extinction is part of a natural cycle. Nobody gets to stick around forever. It's estimated that more than 99 per cent of all species that have ever lived on Earth have gone extinct, and it's likely that, given enough time, all of the species living today would eventually come to a natural end too, to be replaced by others better suited to the world in which they find themselves. That's how evolution works. Scientists who study zoology, ecology and other related areas will expect a certain rate of extinctions in any habitat, as new species evolve to fit the demands of their environment and others die out.

While it's a process that's going on all the time, it's not always slow and gradual; there have been times when a cataclysm has struck, wiping out huge swathes of life over a relatively short period. The single biggest

cause of mass extinctions has been abrupt changes in the Earth's carbon cycle, often as a result of volcanic eruptions that released gases such as carbon dioxide into the air over centuries, causing extreme global warming, changing habitats and starving the seas of oxygen. There have been at least five of these events that we know of in the last 500 million years, including the one that wiped out the dinosaurs.

These mass extinctions are dramatic marks on the fossil record, but each one was a geological event that occurred over thousands of years. Even without climate change, human activity has already caused a steep increase in the rate of extinctions in the world today. They're happening at hundreds of times the natural rate, and much of it has nothing to do with climate change; they are instead the result of hunting, overfishing and the destruction of habitats. But like the bushfires in Australia, or the bleaching of the coral, climate change will impose an increasing price for the extra carbon dioxide we're pumping into our atmosphere. For every disaster we suffer, the animals and plants who live in ever-shrinking habitats will suffer worse. We are reaching the point where we're witnessing a mass extinction on the scale of those the Earth has experienced before – but *humans* haven't. The loss of species that we're seeing today is estimated to be between one and ten thousand times higher than the natural extinction rate.

Looking on the bright side

Having discussed weather disasters, geological cataclysms and mass extinctions, this may seem like a strange place to start offering hope, but that's what I want to do. So, let's start from where we are.

This problem is of our own making. Climate change is not being imposed upon us by some external power. It's bearing down on us, not because of a random asteroid strike, or a series of volcanic eruptions, but because *we're* burning fuels that spew out carbon dioxide, using fertilisers that emit nitrous oxide and farming livestock that belch methane. It is a complicated problem, and hard to understand, which is why we need experts to study all the different aspects of it and tell us what to do – and we need to listen to them.

Up until now, that's been one of the single biggest problems: the people in power are not listening enough to the experts who are warning us that our actions are harming our world. And yet, there are many parts of this that none of us need to have explained to us.

Let's start with animals, because we can relate to animals. Our relationship with them is, well ... complicated. We have pets that we treat as part of our families, we live on or near farms, we watch videos of animals, we enjoy stories about animals. We also eat animals – they're an important part of our food chain. Interacting with animals reminds us that we are still animals ourselves, that their environment is *our* environment. We feel a tangible connection to them. And we need all the same things as they do to survive.

Remember that in many cases it's not climate change but direct human activity that's increasing the rate of extinctions.

If this is something we're doing, then we can *stop* doing it. Humans are not helpless. It is in our power to reduce the rate of extinctions in the wild. Most of us – even those of us who eat meat – do not want animals to be hurt unnecessarily. We can take an interest, demand to know how and why these extinctions are happening and we can work out how to stop it. There are already people who know what to do, who have studied and campaigned on this. We need to listen to them. There are already laws in countries all over the world that have been passed to protect wildlife. We need more of them.

The main cause of extinctions is loss of habitat. There are already laws in every country in the world that govern who gets to claim land for farming, or construct a new building or dig a new mine or build a dam across a river or cut down a forest. There is a limited amount of land in the world, and the way it's used is mostly controlled by governments. In some cases, the laws are there to protect the environment, but more often, they're used to govern how new developments of any kind will affect other people. These laws do not always work, or are used in destructive ways. Money is normally involved.

But it's *humans* who are in charge. We can decide to do more to protect the environment, to protect wild habitats. We can make it part of how we develop our civilisation from this point onwards, not just for the good of plants and animals, but as part of living more in tune with the natural world. We don't need to keep destroying habitats. Humans control this, so humans can stop it from happening.

This will *not* stop climate change. It will, however, start making us more aware of all the ways in which our own lives are connected with, and dependent on, the environment we all share. Maybe if we can save all those animals and the wilderness they inhabit, we're capable of saving ourselves too. Because if we let nature start claiming back some territory, *nature will help us*.

Unstoppable growth

Up until now, our civilisation has consumed everything nature had to offer. It was there, so we took it. You could argue that it's the reason we were able to build a civilisation in the first place. We killed and ate and built and grew and now here we are, the mightiest creatures ever to live on this planet. We've been unstoppable. And that civilisation has allowed us to become educated and enlightened and now we *see* – we see the price our world has paid for this empire we've created.

I don't want people to feel guilty about this. No one person is responsible for where we find ourselves, and I don't think guilt is a good way to motivate anyone; it makes you miserable, and it's hard to get moving on anything when you're miserable. And we do need to motivate people. Because this is our world, it is the only one we have and we can't ignore what our own enlightenment has shown us.

Our entire species has one home, we inhabit a small blue dot in the vastness of space, and even then, though it is beautiful and awe-inspiring, it's only a thin delicate skin around a ball of rock that keeps us alive. We have to start taking better care of our thin layer of atmosphere because we rely on it for *everything*.

Fortunately, the same environment we've taken advantage of for so long can, once again, become our greatest resource in slowing down – and adapting to – our changing climate. It's not enough to stop damaging it, however; we have to step right back and let it do its thing.

Left alone, nature will consume carbon dioxide and make more oxygen, and it will do it on a worldwide scale – which is exactly what we need it to do. Just like us, it wants to grow and keep growing. It can start claiming back its lost ground; we just need to *get out of its way*, and give it some space. Sure, there's plenty we can do to help it along, like planting trees or reintroducing species that have disappeared, but nature's appetite for success is as great as ours, and it will fill any gaps that we leave for it. It will recreate the kind of biodiversity that makes ecosystems thrive.

This has been done successfully all over the world – it has even happened completely by accident.

The Demilitarised Zone, or DMZ, is a border area that runs west to east between the nations of North and South Korea. After the Korean War, which took place between 1950 and 1953, the countries agreed an uneasy peace, but each was certain that the other would try to invade. So they left this strip of land between their borders, which is about four kilometres wide and 250 kilometres long, littered it with landmines – the type of bomb that explodes when you step on it – and surrounded the whole place with barbed wire, watch-towers and machine-gun nests. Trying to cross the DMZ without authorisation is an excellent way to get yourself killed, so hardly anyone does. The area was originally farmland; with the humans gone, however, nature took over.

Now, South Korea's Ministry of the Environment claims that there are five thousand species of plants and animals in the area, including a hundred that are protected. Nobody's entirely certain of the number because you'd have to risk getting blown up to find out for sure. It's ironic that a place scattered with landmines and machine-guns offers more hope for wildlife than land cultivated for farming but, to put it bluntly, we need to look at what else humans are doing to harm our world, and put a stop to it.

Guilt is not a good motivator, but *expectation* is. If you expect something to happen, you're more likely to *make* it happen, and to encourage others around you to do the same. The wild world is not separate from us; it is not a different environment to the one *we* inhabit. It is the foundation of all life on our planet, and we should expect our political leaders and businesses and communities to work together to protect it ... to protect *us*. Humans have an influence over our leaders, and we humans should expect the very best treatment for our world.

Once we've accepted that nature's problems are *our* problems, and helping nature is the first step to saving ourselves, then we can start working out what else we have to do. That will come down to what each of us feels most passionate about, and where our individual strengths lie. Nobody can do everything, but everyone can do *something*.

And along with the air we breathe, our most direct connection with our environment – and our greatest influence on it – is through the parts of it that we eat and drink.

Friends of the Earth Ireland

Biodiversity is important for maintaining a safe and healthy climate. We can help nature by planting pollinator-friendly flowers in our gardens and parks. Indigenous peoples protect and care for about 80 per cent of the world's biodiversity. We can learn from indigenous peoples' cultures and how they identify as Earth guardians. They respect and value the earth; they don't see resources and species as commodities but as crucial elements of our life-support system.

CHAPTER 12
THE NEED TO EAT

Farming provides us with our food. From the earliest days of keeping animals and then planting crops, farming has enabled us to create the advanced society we have today, because it was hard to get anything else done until we'd made sure we kept on living. Despite the vast amount of damage it has done to our wild environment, it's a hard fact that we need to farm a large part of the Earth to feed our population.

Ireland is no exception, and we are unusual in Europe in that farming produces about a *third* of all our country's greenhouse gases – in Britain it's more like 9 per cent. While carbon dioxide is the most famous greenhouse gas, making up 76 per cent of global emissions, it's not the only A-list celebrity, and it doesn't play the main villain in the world of agriculture. Those roles are played by methane, from the belches and manure of animals, particularly cows, and nitrous oxide, from fertiliser. While there is far less of these gases in the atmosphere, and they don't hang around as long as CO_2, they pack a much bigger punch. Methane can be twenty-five times as potent as carbon dioxide, and stays in the atmosphere for more than a decade. Nitrous oxide can be about *three hundred times* more potent than CO_2, and can stay in the air for a hundred years. Carbon dioxide, though it's a relative wimp in comparison to its beefy co-stars, hangs around for *centuries*.

Biological warfare

Just as peat, coal and oil develop from the remains of living things, so does soil, the basis for all farming. Soil is a limited resource; it has to be maintained and nourished or it can become barren, where nothing can be grown in it. In order to work the land, a farmer must have an in-depth understanding of it, and though farming is now carried out on an industrial scale, supported by science and sophisticated processing systems, the same awareness of the natural world that was needed ten thousand years ago is still needed today. We have to grow things to eat, but that same soil is also loaded with carbon and is a source of greenhouse gas emissions, and we have to figure out how to keep feeding ourselves without destroying our environment in the process. And while farming produces a large proportion of greenhouse gases, it is also one of the most vulnerable victims of climate change – a situation that will become progressively worse. And anything that hurts *farming*, hurts *all of us*. Because it's where we get our food.

One of the biggest issues that climate change has created lies in food security. Even small changes in average temperatures can have big effects on agriculture, which relies on predictable growing seasons. Most commercial crops don't adapt easily; having been bred to grow in certain climates, they will struggle as our weather changes, which will mean lower yields – less food. Extreme weather events like droughts, wildfires and floods will destroy crops.

Rising temperatures will bring new enemies; invasive species of insects, as well as bacteria and fungi that will feed on our food before we can. We know that the conditions we're creating will cause the destruction of our crops – we're already seeing the consequences – so it's vital that we take action to reduce the damage and adapt our farming to the changes coming down the line.

Many people already know that livestock – mostly cattle and sheep – are belching out methane at an industrial level. The human appetite for meat and dairy products is one of the most damaging contributors to climate change. It depresses me to say this, as I love both – but I also know I don't have to consume them all the time. Scientists are researching techniques to change cattle feed to reduce their methane emissions, such as adding seaweed to their feed, or using protein-rich concentrates rather than their usual grass diets, though cattle still need grass to stay healthy.

However, the most effective way to reduce the emissions from our livestock is to reduce the number of cattle and sheep. If we consume less meat and dairy, we won't need so many of these animals, and it will leave more room for growing crops or even letting more land grow wild again. Unsurprisingly, many livestock farmers are not great fans of this idea.

In large industrial farms, 'monoculture' is common; this is where the same crops are grown on the same land every year, crops such as wheat, soy, rice, palm oil or coffee. When you're farming on a big scale, it's more efficient, because you plant, maintain and harvest all your crops in the same way. However, this type of agriculture is highly intensive and drains the soil of nutrients, making it harder to grow anything the following year. Farmers compensate for this by using nitrogen fertilisers, a process that releases nitrous oxide – a greenhouse gas that is hundreds of times more potent than carbon dioxide– and can result in those fertilisers ending up in the rivers near by. It can also increase the risks caused by pests or disease; if you have huge areas of one crop and one part is attacked by something that has a particular taste for that crop, you could lose the whole lot of it. To try and prevent this, farmers use pesticides, which in turn can lead to more pollution. Using pesticides can also kill all the useful insects along with the ones doing the damage. We really need those insects to pollinate our food crops, help keep our soils healthy and provide food for other animals.

On top of all this, increased specialisation means that a farm concentrating on one type of food has to distribute it more widely to reach a bigger market. That means more transport, which means burning more fuel, adding even more greenhouse gases to the atmosphere.

Different geography and different soils can demand different types of farming, so something that works in one region might not work somewhere else. Like plants and animals in the wild, farming has to fit its environment. But even something as basic as ploughing soil will release carbon dioxide. Imagine that – you don't even have to burn it, you just have to stir it up. Research has found that soil that isn't ploughed before planting releases much less carbon dioxide and suffers less soil erosion. Just reducing ploughing could help keep carbon in the ground.

Custodians of the land

While farmers live with the land and work with it, each farm is also a business that must serve its customers in order to survive and succeed. This means that every farming business has to balance the needs of the land on which its existence depends with the need to make as much money as possible out of that land. Any change to the way they do things involves a cost, and we cannot ask farmers to make changes that will put them out of business.

At the moment, farms in most developed countries receive a lot of financial support from governments for certain types of farming – it's a recognition that what they do is vital to our society. Everyone paying taxes helps to pay for farms. Our governments could offer support for *changing* the way we farm, switching to what's sometimes called 'regenerative' farming. If farmers could be encouraged to stop producing monoculture crops, to reduce livestock farming, to mix livestock and crops in a way that increases biodiversity, to reduce ploughing, to plant trees and to allow some of their land to grow wild, these measures would go a long way towards tackling climate change. It would reduce greenhouse gas emissions and make our farming more sustainable.

Again, it comes back to expectation. We should expect that farming contributes to the health of our environment, rather than taking from it – and part of that responsibility lies with, well ... everyone who *eats*. Food does not just appear on supermarket shelves. We should show more appreciation of, and interest in, where our food comes from. We should educate ourselves about how it's produced, who produces it and how far it has travelled.

Our food is one of our most direct links with our environment, and it's farming that connects us to it.

The giver of life

When the scientists who search for life on other planets gaze out into space, the key thing they look for is the presence of liquid water. When life began on Earth, billions of years ago, it began in the seas, even before there was oxygen in the atmosphere. And it is still essential for all life today, humans included. If our civilisation has one task that is more important than farming, it is to maintain a safe supply of drinking water.

The adult human body is about 60 per cent water, and it needs to stay that way or it will die. A human baby is about 75 per cent, which might explain why they're always leaking. Water is a vital component of our cells: it transports nutrients and oxygen as part of our blood and regulates our temperature through sweating. It's a shock absorber for our brain and spine and for a foetus in the womb. It keeps our joints lubricated and forms saliva in our mouths. It helps with digestion, and to flush out waste products as part of our poo and pee.

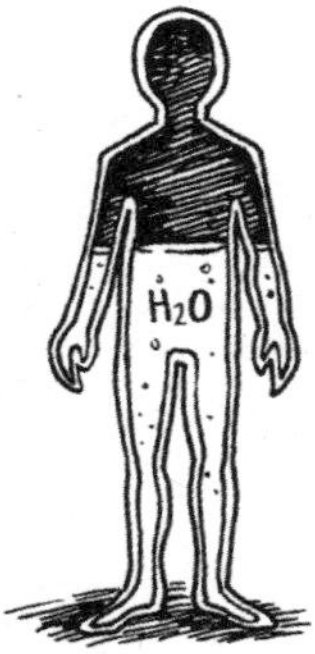

The average adult needs to consume between two and three litres of water a day, though we don't actually have to drink that much, as we also get water from our food. We cannot survive for more than a few days without water. It's that important. We saw earlier how water weaves its way around the globe; how evaporation from the oceans travels as water vapour through the atmosphere and falls as rain or snow. When the rain falls on land, it finds its way back to the oceans through streams and rivers. But it's worth noting that the vast majority of the water in the world right now is undrinkable.

That's because 96 per cent of it is salt water, in the oceans, and nearly 70 per cent of the total amount of fresh water is locked up in ice and glaciers. Another 30 per cent of fresh water is in the ground, which we have limited access to through wells. Our main sources of drinking water, however, are rivers and lakes, which only make up about 93,100 cubic kilometres. That's about 1/150th of 1 per cent of the total water on the planet. It seems like it should be more, doesn't it?

And of all the water humans use, 80–90 per cent is used, not for drinking, but for growing crops. So not only is water vital for stopping us from dying of thirst, it's kind of important for stopping us dying of starvation too. That's the Earth for you – *everything's* connected.

Because food and water and the land that provides them are so important, nations put a lot of effort into keeping control of them. And when there's a shortage of these things, well ... that's when we start fighting over them.

Friends of the Earth Ireland

Growing herbs and vegetables is a fun and easy way to help the environment while helping us to reconnect with nature and where our food comes from. Composting is a great way to use food scraps to make chemical-free fertiliser. Buying local food reduces air-miles and eating less animal produce reduces pollution.

CHAPTER 13
CONTROL OF SUPPLY

Of all the things that climate change is going to affect, it's our water supplies that are going to be hit hardest. Some of these effects are obvious. If there's less rainfall, there'll be a drought, so there'll be less drinking water. If there's a heatwave, people will use more water, and water will evaporate faster, so there'll be less drinking water. If there's a drastic drop in temperatures, pipes can freeze, and possibly burst, and there'll be less drinking water.

And if an area suffers serious flooding, there could be less drinking water.

That might seem weird, because if there are floods, then surely there is ... *lots of water.* However, flooding can often bring problems with water supplies. Storm surges can contaminate ground water with sea water. Heavy rainfall can wash fertiliser off fields into rivers, which can poison the water and also wash downstream to the sea where they can cause those algae blooms which result in dead zones. Sewage systems can overflow, so flood waters have to be treated as contaminated, as they could be carrying bacteria or toxic chemicals. The risk of water-borne diseases like cholera increases.

For societies that get most of

their water untreated, straight from rivers, lakes and wells, flooding can be catastrophic, but even in wealthy countries where water treatment plants are the norm, they can face major problems.

After Hurricane Katrina hit America's Gulf Coast, the residents were faced with a lack of drinking water and sewage treatment, and no electricity. There were swarms of insects breeding in the flood water, and the poisoned waters had tainted fish and shellfish populations. Because of the oil refineries and chemical plants along the coast, the residents also had to deal with major chemical spills and other sources of hazardous waste. And this was all in the richest country in the world.

For creatures that can't go more than a few days without clean water, extreme weather events are rarely a reason to celebrate.

Surface tension

Control over water has always been a key element in controlling the environment – and claiming territory. It could be two farmers arguing over who has the rights to a stream to irrigate their crops. Or it could be three nations facing off over who has rights to the water in the longest river in Africa, and possibly the world.

At the time of writing this book, negotiations were still going on between Ethiopia, Sudan and Egypt over the construction of Africa's largest hydro-electric dam. Tensions were rising. The Nile rises in Ethiopia, flows north through Sudan and then Egypt, where it reaches the Mediterranean Sea. Ethiopia is building the Renaissance Dam to generate electricity, but it will also help regulate the flow of the river and reduce the risk of flooding – in theory. But a project this size will cause major environmental damage, and Ethiopia wants to take five years to fill the reservoir behind the dam, during which time less water will flow through Sudan and Egypt.

Egypt in particular is objecting to this, and wants Ethiopia to take more time and less water, stretching it out for seven years. Since *90 per cent of Egypt's fresh water* comes from the Nile, its leaders are worried that the reduced flow will put millions of farmers out of work and affect drinking water and food supplies. However, because Ethiopia controls the top of the Nile, there may not be much that Egypt or Sudan can do about it. The dispute between those three countries is a political problem, and will be resolved in a political way. Despite what Hollywood would have us believe, many of the world's worst problems are solved slowly, by the right people sitting down and having a chat, rather than indulging in gun-fights or a liberal use of explosives.

The conversation is only the start, of course, but the more people who agree on a course of action, the more likely it is to get done. The United Nations was founded to tackle challenges on a global scale, and water is high on its list of priorities. It estimates that there are about two billion people living in countries where water shortages are a real problem. Water shortages will also mean food shortages. The Global Water Institute estimates that 700 million people could have to leave their homes because of issues related to water shortages by 2030.

That's a *lot* of people on the move – people who may have been forced to leave in desperate circumstances, who can bring little or nothing with them.

It's rarely as simple as just running out of water. When you're struggling with something as basic as your water supply, other issues can build up on top of that over a long time. Without water, it's impossible to farm. With no farming, whole populations can go poor and hungry. It becomes difficult to keep order. Poverty and hunger drive people into desperate situations; ask yourself what *you'd* do to keep your family alive. Desperation increases the risk of violence. And conflict, in turn, can cause water shortages, so it just makes everything worse.

War and water

Water can be involved in a conflict in three different ways. First, it can be the *cause* of fights: this happens when there is a lack of water. Secondly, it can be used as a threat or a weapon: depriving populations of water is a powerful method of warfare. And, thirdly, – and this is the really stupid bit – water can be a *casualty* of conflict. This happens when water supplies are contaminated or water infrastructure (wells, pipes and so on) is damaged as a result of a conflict. That's humans for you: a reliable water supply might be the most important thing in our civilisation, but we're not averse to blowing it up if it stops our *enemies* from having it.

Modern infrastructure is no guarantee of a water supply – think of the aftermath of Hurricane Katrina in

the United States. Another example would be Aleppo, in Syria. Before the civil war in that country, Aleppo could boast architecture that ranged from ancient to state-of-the-art. One of the oldest cities in the world, it had the kinds of industries, shopping centres, public services, office buildings, universities and transport systems that you'd find in any major city around the world. The region was already suffering a prolonged drought when, in February 2017, the United Nations reported that Aleppo had had its water supply cut off as the city was pounded in another series of battles in the years-long war. Nearly *two million people* had no drinking water.

At different points, this 'civil' war has involved the United States, Russia, Turkey, Iran, Saudi Arabia and Israel, among others. With large parts of the city in ruins, and violence erupting all around them, innocent civilians were completely dependent on the United Nations, along with partners like the International

Committee of the Red Cross and the Syrian Arab Red Crescent, to truck in supplies of water and provide fuel for the pumps in wells that drew up ground water. Because of this conflict, people in a modern city were at risk of dying of thirst.

Given that the most important task civilisation has is to ensure people have a good supply of drinking water, having nearly two million people going without drinking water would suggest we're not doing this civilisation thing very well. Clearly, we should be trying harder.

Holding on to what we've got

Knowing that climate change is going to disrupt and threaten water supplies, and knowing that most of the drinkable water we use goes towards growing crops, it's ironic that, worldwide, the single greatest polluter of drinking water is ... wait for it ... agriculture. Run-off from fields can result in fertilisers, pesticides, herbicides, bacteria and parasites in rivers and streams. The single biggest type of pollutant is nitrates, from nitrogen fertilisers or manures. These can have dangerous health effects, and also damage ecosystems.

Many countries have regulations to govern the levels of nitrates in water sources, and even then, those levels are damaging the soil and water we rely on for survival.

The second biggest polluter is the textiles industry. The companies that make our clothes. Because we all need clothes – and we love to look *fabulous*. Some of the chemicals used in dying fabrics in particular can be highly toxic, and even when diluted in rivers and streams, they can build up in the bodies of the organisms that live there and end up in the fish that we eat – as well as poisoning everything else. Non-organic fabrics like polyester give off plastic micro-fibres every time they're washed, which in turn end up in increasing concentrations in the environment.

Mining is also a major cause of pollution. Some of it can be released from the excavated rock itself, such as arsenic, cobalt, copper, cadmium, lead, silver and zinc, which can be carried downstream as water washes over the rock surface. Some comes from substances such as cyanide or sulphuric acid, which are used in the mining process to separate the mineral from the ore.

We can only access about 1/150th of 1 per cent of the total water on the planet to drink and supply our crops – and all the ecosystems around us. We can't afford to be *poisoning* any of it. And while some of the poorest countries will be hit worst by water shortages, the most powerful country in the world is not immune either. Since around the start of the century, the Colorado River, one of the longest rivers in the United States, the mighty flow that carved out the Grand Canyon, has seen a drop in water levels. Less snow up in the Rocky Mountains, where it begins, means the glaciers that feed all the rivers there are shrinking. It is a water supply that supports millions of lives.

Behind the world-famous, towering structure of the Hoover Dam lies Lake Mead, the largest reservoir in the United States, which supplies countless farms in several states, as well as major cities such as Phoenix, Los Angeles and Las Vegas. Lake Mead is supplying more water than it's taking in from the Colorado River, and its levels are approaching a point where the water supply might fail altogether. It is *forty metres lower* than it was twenty years ago. 'Bathtub rings' mark the rocky sides of the lake, showing where the original levels were.

Las Vegas, built in the Nevada desert, is a city famous for its big-spending party animals, huge hotels and casinos, fountains and golf courses. They have been rationing water for years, and have had to become extremely inventive in ways to save water, yet it may still not be enough. Climate change is expected to cause ever decreasing snowfall in the Rockies. The glaciers will keep shrinking, which will mean the great Colorado River will continue to decline.

That's without any one-off environmental disaster, any huge polluting event or civil war. It's just because there's less snow falling in the mountains.

Fortunately, we've had a bit of experience with this. Water supply is a problem that humans have been solving ever since we figured out how to start carrying some with us, instead of having to find a stream every time we were thirsty. For every place where there's a shortage of water, there are people coming up with solutions.

We can start with the obvious: we need to protect what we have. We can increase the amount of drinking water by *not polluting our sources of water.* Most countries already have laws in place for this, and they could be stricter and better enforced. We can support the ecosystems that help keep our waterways healthy. There are plenty of technological solutions being developed too, including adapting buildings to use grey water (waste water from sinks, baths showers, dishwashers etc. – but *not toilets*) for flushing toilets. Sea water can be used for this too, for locations on the coast.

In places like Pakistan, Yemen and Ethiopia, landscaping with stone spillways is being used to direct and control water in farming areas prone to flooding, storing it for use on the farms. In the Philippines, mobile toilets that don't need to be plumbed in are being tested. They're designed so that, when they are emptied, the waste can be processed into fertiliser and grey water. There are even systems now that draw moisture from the air using solar power. They can produce water in a desert.

There are so many solutions, and smart, imaginative people are creating more all the time.

Friends of the Earth Ireland

The Keystone XL pipeline was planned to bring oil from tar sands in Canada through the US, along territories of numerous indigenous tribes, to the Gulf coast. Resistance first came from indigenous people protecting the land, but soon many more joined the fight against the pipe. If the pipe was built, it would have added to climate pollution and contaminated water supply. This was such a big campaign, President Biden cancelled the pipe on his first day in office in 2021.

CHAPTER 14
POWER STRUGGLES

Most of our fossil fuels are burnt to generate power in one form or another, and as well as being vital for drinking and food production, water is also one of the best alternative power sources available to us – another reason to take good care of it (as if we needed another reason). The Hoover Dam, which holds back the Colorado River, is an enormous construction. One of the world's largest hydro-electric dams, it's such a dramatic location that its soaring, curving, cliff-like face has featured in numerous Hollywood movies. And yet it is eclipsed by the Three Gorges Dam, which spans the Yangtze River in China.

About 181 meters high and 2,335 metres long, this dam creates the Three Gorges Reservoir, which has a surface area of about 1045 square kilometres and stretches about six hundred kilometres upstream from the dam. Which makes it approximately *five times* larger than the Hoover Dam complex – and it generates *eleven times* more power. And that's still only a small portion of the nation's needs.

China today is our modern world taken to the extreme. The country has become a manufacturing powerhouse, with mega-cities growing up around its industrial complexes. All that productivity has come at a price. Its coal-fired power stations and its massive steel production plants have turned the air to a choking soup over some of the country's major cities. China has become the world's largest emitter of climate-warming greenhouse gases – though the United States and Europe have still contributed more overall – and it's estimated that air pollution kills nearly a million of its people every year. It's important to remember that a lot of the pollution that China is enduring is due to the fact that they manufacture so many products that are used in the western world. Our consumer lifestyles are contributing in a big way to their health problems.

The extraordinary levels of pollution in China have led to episodes of what some call 'airpocalypse', where the local governments sometimes have to issue orders for people to stay in their homes because the smog is so thick and toxic, it's dangerous to go outside. Traffic is restricted, schools are shut and flights are grounded. Venture out in the streets and you risk inhaling particles that not only can cause breathing problems, but prolonged exposure can also lead to heart attacks, strokes and neurological problems. Needless to say, the millions of human beings who drive China's success are less than happy with the lack of breathable air.

However, the Chinese government is intent on doing something about it. They have a culture of long-term planning and recognise the threat of climate change. In many areas of technology, they are leading the charge to get free of fossil fuels, including a plan to close their coal-fired power stations. Instead, they've started building vast arrays of solar panels and other large-scale infrastructure projects – like the Three Gorges Dam.

As a means of replacing fossil fuels as a power source, the dam is an incredible achievement. Its thirty-four huge generators provide energy to millions of homes and businesses – 22,500 megawatts, the equivalent of *twenty* coal-fired power stations. Hydropower is now China's second largest source of electricity after coal.

But the dam has been controversial from the very beginning because ... well, quite frankly, it's an absolute monster. It cost $37 billion to build, and when the reservoir filled, it covered thirteen cities, 140 towns, and more than 1600 villages. About 1.3 million people were forced to leave their homes. Farmland and ecosystems were lost on the same scale. The reduced water flow along the Yangtze River has already caused water shortages in central and eastern China, including in the city of Shanghai. Ships have ended up stranded because of the low water levels.

And yet, the claims that it would help control flooding proved somewhat ironic when, in the summer of 2020, that area of China saw months of torrential rainfall, which led to the highest flood water in recorded history. Here was another one of those extreme weather events that are increasing due to our changing climate, and it showed that the dam's ability to manage all that extra water didn't match the ambition of its architects.

The raising and lowering of the water level in the reservoir also destabilises the land around it. Water seeps into the soil in the cliffs surrounding the reservoir, causing enough erosion to make the ground slip, resulting in landslides and mudslides. The site also sits on two major fault lines: Jiuwanxi and Zigui-Badong. Scientists fear that the dam is so massive, its huge changes in water pressure as the reservoir levels change could actually trigger earthquakes.

The Three Gorges Dam is an incredible source of sustainable energy, and yet there are serious concerns about what the cost will be to the environment around it. When it comes to the new infrastructure we are going to need all over the world to tackle the challenges of climate change, this project has raised questions that will demand some tough and even uncomfortable answers.

There's no doubt that giant infrastructure projects like this will have to be part of our response to climate change, but in most places, they simply won't be possible, and may not be the best solution anyway. China is not a democracy, so the people cannot vote their leaders out of office. If the government decides to build a colossal dam and move nearly 1.5 million people out of the way, they can do it. This is not the case for most of the industrially developed countries in the world – the ones that create the most greenhouse gases and who have the greatest responsibility to change their ways.

And they have their work cut out for them, because 40 per cent of the world's energy comes from one type of fossil fuel. We are still far too dependent on coal-fired power stations.

Awakening something ancient

Earlier in this book, I talked about coal, that hard black concentrated form of carbon that burns so wonderfully hot, and has contributed so much to the growth of our civilisation. Created over millions of years, under huge pressures, it is a rich and potent source of energy. And, unfortunately, of carbon – carbon that has been stored in the ground, far from the surface, for all that time.

A power plant burns coal to heat water to create steam, which drives turbines, which generate electricity. There are a lot of advantages to using coal as a fuel. In its most basic form, it is a simple brute force form of energy. It's very old technology, so we know it works, and we can build these kinds of plants relatively easily. A nuclear power plant, by comparison, is much more powerful, but vastly more complicated and difficult to build.

With coal, we can turn the fire on and off at will, unlike sustainable sources like wind or solar, which effectively last for ever and are free, but provide an unreliable supply. You can build a coal-fired power plant pretty much anywhere, unlike a hydro-electric plant, which needs a major source of running water. As a fuel, coal is unhealthy for us and for the environment, but it's cheap, energy-rich and convenient too, and like junk food, that's why we *love* it.

A one-thousand-megawatt coal power plant can burn over eight thousand tonnes of coal per day. That's a lot of carbon dioxide that wouldn't otherwise find its

way into the air. This is the environmental equivalent of creating a dinosaur from ancient DNA and letting it loose on the world. The largest coal-fired power plant in the world, the Datang Tuoketuo power station in the Inner Mongolia Autonomous Region in China, uses nearly twenty million tonnes of coal per year.

But, though coal has been the biggest villain in this story, it's not the only one.

Some older power stations run on oil, though they are already being phased out. They are normally only used to top up the supply when a little extra power is needed on a network, as they are so expensive to run.

For a long time in Ireland, peat (or turf, as it's known to the Irish) has been used as fuel for our power stations – a practice we share with countries like Finland, Indonesia, the Russian Federation and Sweden. It's a really dirty fuel that has not only caused pollution, but also meant strip-mining bogs that drew huge quantities of carbon dioxide from the air, making things even worse. As the country faces major pressure to reduce our emissions, and to protect those delicate ecosystems, Bord na Móna, the company responsible, which is owned by the Irish government, announced in 2021 that it was ceasing all harvesting of those precious bogs. Within a few years, the final stocks will run out, and Ireland will stop all large-scale burning of this fuel and the terrible environmental damage it causes.

No matter what fossil fuel we're talking about, however, we're quickly using up a resource it took millions of years to make, one that we can't replace – and if we keep using what's left, it will threaten our very civilisation. We are playing *Jurassic World* with our atmosphere, taking something long dead and releasing it into our modern world, without giving enough thought to the consequences.

The end of coal – a future of gas?

Even the companies that *build* coal-fired power stations are finally admitting that the end is nigh. It's becoming a liability, and it's starting to show in their profits. In 2020, Samsung C&T Corporation and the Toshiba Corporation, two of the biggest power plant manufacturers, both announced that, after they'd completed their current projects, they would no longer be building these kinds of power stations. That's how unpopular coal has become. People just *don't want it* any more.

That's not to say that these plants are being replaced with sustainable sources of energy. As the fossil-fuel industry gives up on coal, it's promoting natural gas as a replacement. Worldwide, it's the second most popular fuel for generating electricity. The use of this fuel is expanding fast, with plans for thousands of gas-fired power stations to be built over the next decade. This, in turn, will create a great demand for gas, and will commit us to drilling for more – including fracked gas.

'Fracking' is short for 'hydraulic fracturing'. To hear it described, it sounds like something a cartoon super-villain would do to destroy some innocent, peace-loving community. A shaft is drilled deep into sedimentary rock, such as shale, and then a fluid made up of water, sand and chemicals is pumped into the hole at extremely high pressures. This opens networks of cracks in the rock so that the gas can be pumped out. The process is incredibly destructive to the environment – the pumping in of the water can actually cause *earth tremors* – and it is already banned in Germany, France, Ireland and the Netherlands, among others. Apart from the damage caused to the rock, many of the chemicals used are highly toxic and can contaminate water supplies. There we go, poisoning our water again.

Gas may seem like an improvement on coal, emitting half as much carbon dioxide, but that's like saying it's better to smoke *ten* cigarettes a day than it is to smoke *twenty*. And building new gas-powered stations now means that we'll continue to use gas as a fuel for decades into the future, rather than committing completely to renewable sources of energy.

This kind of change is too little, too late, and it is now a major focus for environmentalists to stop fossil fuel producers from feeding our addiction to these fuels and, instead, to end our reliance on them altogether.

Our hunger for power

Try and imagine your life without electricity. I don't mean for a few hours or a day or two, I mean with *no electricity, ever*. Imagine all the normal, everyday things you couldn't do. It is such an integral part of modern life that we can almost forget what a wonder it is. There are places all over the globe that still have little or no electrical supply, but here in Europe, it's taken entirely for granted. And we keep finding new ways to use it.

Whether it's charging your phone, boiling the kettle or lighting up a football stadium, that's a lot of electrical power when you're looking at an entire continent. So where does it come from?

Well, in the EU, renewable power produced *more electricity than fossil fuels* for the first time in 2020, a total of 40 per cent, compared to 34 per cent for fossil fuels. Most of that came from wind turbines and solar panels, but it also included hydro-electricity and bio-energy (burning things like wood pellets and other plant products). Denmark was the overall leader, followed by Ireland and then Germany. Most of the remaining 26 per cent of electricity was accounted for by nuclear power.

Modern society's demand for power has been rising steadily, and as it has grown, each country has responded in a different way, depending on what suited them best. A generation ago, the idea of 'power' centred around a giant complex owned by a large company, consuming huge quantities of fuel and feeding electricity out across the country. Because of the scale needed, we still tend to think along those lines. A major energy company finds a big space, builds some giant turbines, and burns coal or gas to make steam. Nuclear power is a highly refined and incredibly potent version of the same idea – though one that *doesn't* produce greenhouse gases as part of its waste.

As we've become more aware of the damage we're doing, however, and more committed to leaving fossil fuels behind, we've steered our future towards sustainable sources of energy, and our environment has started having a greater influence on how and where we generate our power. Just as each region's geography affects its weather, it also affects the nature of the power available to its inhabitants. Our search for sustainable energy is making us pay more attention to our natural resources, and it's forcing us to adapt our technology to our landscape. We've already made a lot of progress, and things are improving all the time.

The wind and the sun

Ireland is a small island with a moderate climate and a lot of rain. We have plenty of rivers, so we can harness them for power, and we do get our fair share of wind, especially on our coasts. We already have three large hydro-power stations on dams, though our small, low-lying rivers don't lend themselves to the scale of projects in the United States, China or even places like Scandinavia. We haven't cracked tidal power yet, and we may be decades from it, because the machinery

to produce that has to be built *under the sea*, which eventually breaks any machine we leave in there for a long time. But it has potential, and if we could get it to work, we'd have a free source of reliable and never-ending power. Unlike the sun and wind, the tides are always moving against our coasts.

There are small geothermal installations dotted around the country – piping heat up from the ground – but because we don't have much in the way of volcanic activity, these are mostly used to help heat individual homes or other buildings. It isn't a realistic option for generating electricity on any scale, unlike in places like Indonesia or Iceland. For the foreseeable future, our main sources of sustainable energy will be wind and solar, and when you have a lot of coastline, offshore wind offers even greater opportunities.

A turbine is a kind of wheel that generates electricity when it spins, and that motion is driven by a gas or liquid moving past it. If you've ever seen an old mill wheel in a river – the type that was used to turn a mill stone to grind grain – a turbine is like a modern version of that, spinning up electricity instead of turning a big stone. In fossil-fuel power stations, coal, gas, oil or peat are used to heat water to create steam, which is forced through pipes at high pressure to drive turbines. Nuclear plants effectively work the same way. Wind turbines, on the other hand, are more like the old windmills, using their giant, propellor-like blades to catch the wind and create electricity that way.

There are two basic types of solar panel: one that can charge your phone, and one that can't. A solar thermal collector doesn't generate electricity – it's a kind of sun-powered kettle. It uses either a material like copper, aluminium or steel or an array of vacuum tubes to soak up heat from the sun and transfers that heat into a tank of water.

A photovoltaic, or PV, solar panel makes electricity out of sunlight. It's made of wafers of silicon, a crystalline material that's one of the main components of glass and is used in making computer chips. These wafers are arranged in layers to react with each other, creating an electrical field, a little like a battery. Instead of generating a charge on their own, however, these photovoltaic cells absorb the photons from sunlight and release electrons to produce a source of electricity.

Nearly one-third of all electricity generation in Ireland now comes from wind turbines, and though we still have a lot of work to do to bring our emissions down, Ireland has become a world leader in integrating renewable electricity, particularly wind energy, into the national grid. As far as solar power goes, it may not seem like we get a lot of sunshine, but we actually get 80 per cent of the solar radiance of places like Italy, with the added advantage of regular rain to keep solar panels clear, washing off the dust that you can get building up in more arid environments. Who'd have thought all our rain would be useful for solar power?

It might seem odd, but these two sources of power are something we have in common with the Sahara Desert, where the geography is ... different. As is the size of the place. At nine million square kilometres in area, stretching across several countries in North Africa, what the world's largest hot desert lacks in water, it more than makes up for in sunshine and wind. Large-scale installations of solar panels and wind turbines are already being planned across the desert; so big, in fact, that research is being carried out into how they will change the environment, and even the weather around them.

With global temperatures increasing, the Sahara is spreading, its edges slowly eating up farmland, increasing the risks of food shortages in these areas. However, solar panels absorb sunlight, preventing it from being reflected back into the atmosphere, while wind farms churn warmer air from above into cooler currents lower down. Some researchers believe that, on a large enough scale, these installations could not only generate massive amounts of electricity, they could even bring more rainfall and promote the growth of vegetation.

There we go affecting our weather again.

As photovoltaic cells become ever more efficient and cheaper to produce, solar power has proven to be a growing success story, not just for the environment, but also as a business. In 2020, the International Energy Agency declared solar to be 'the cheapest electricity in history'. We are learning to tap into that same power that's heating our oceans and melting our glaciers. And it has the widest range of applications, from a unit you carry in your bag to charge your phone, to a few panels on the roof of a house, to a vast array, stretching across a desert.

Owning this thing

Fossil-fuel power plants, those giant, smoking polluters we've become so dependent on, are on their way out, but it's a slow process. The problem is that, in most cases, it takes a lot of dams, wind turbines, geothermal heat pumps and solar panels to replace the supply from a single coal-hungry power station, and those installations can still have negative effects on the environment. The bigger the project, the more effect it's likely to have, and before one is built, the impact it could have on local people and their environment has to be investigated.

Whether it's the effect a dam on an Irish river will have on the fish life, or the impact of kilometres of solar panels on a delicate ecosystem in the desert – and yes, even deserts have ecosystems – the benefits have to be weighed against the damage done. The less we interfere with the natural systems we all rely on, the better. On the other hand, nobody wants a power plant near their home either, and we have no choice but to move away from the practices that are pumping greenhouse gases into our climate. From the Three Gorges Dam to a small hydro-power system on your local stream, for every new sustainable project that we introduce into our environment, we need to ask: Will this make things better or worse?

And that's how it should be. The smaller a project is, the more influence people in the area can have over it and the more they can become invested in its operation. Sustainable energy is making the electricity industry more democratic. We no longer have to rely entirely on huge companies for our power. The ability to generate electricity for a single house or a community or a town is becoming steadily more affordable and realistic. And while bigger projects can offer the advantages of scale, any project must still have as little impact as possible on its environment, and those who build them must accept the responsibility that comes with them. That's how we'll keep everything working, by being responsive to the environment at every level, from local to global.

We need to drink, we need to eat, we need space to live and we need to generate energy to do all the things that we do – and that's going to leave a mark on the planet. How big and how damaging a mark that will be is something we can control. If we do things right, the Earth will provide us with all the power we need; clean, constant and secure for millions of years.

But it's going to take us a while to get to that point, and in the meantime, there is one other means of generating electricity. Though it's not very dependent on geography, it cannot be used on a small, local scale, it cannot be produced by small communities, or indeed most poorer *countries* in the world. However, it gives off *no greenhouse gases* in its operation, and is by far the most potent source of energy we've ever had ...

And many environmentalists really, *really* hate it.

Friends of the Earth Ireland

EcoPower is Europe's largest energy cooperative. It started when a community in Belgium got curious about reviving an old watermill – and they got it working to create energy for their community! The more people that joined, the more energy that was created. Today they also have wind and solar that powers fifty thousand homes from 100 per cent renewable energy.

CHAPTER 15
THE LESSER OF TWO EVILS?

Environmentalists tend to argue a lot about the best way to do things, which is only to be expected. Most problems have more than one solution. It's like democracy, but nobody ever gets to be in charge. And while there is plenty of room within the vast topic of climate change to argue about details, nothing is dividing opinion quite like the issue of nuclear power.

Since the 1960s, people have been campaigning against nuclear energy because of the risks it poses, not just in terms of accidents, but also because of the waste it creates. In recent years, some very respected, very vocal environmentalists have been doing the numbers and have changed their minds, deciding that the long-term threat of climate change outweighs the risks posed by nuclear power. That's how seriously they view the climate crisis.

I'm not going to take a position on this, but I want to use this section to look at why so many people think nuclear power is too dangerous, while others believe it's a solution we might have to accept.

A toxic reputation

The most negative images of nuclear power come from the accidents that have created disasters, most famously Three Mile Island in the United States in 1979, Chernobyl in the Ukraine in 1986 and the Fukushima Daiichi plant in Japan in 2011. There were no deaths directly related to the Three Mile Island accident in Pennsylvania, though one of its reactors was destroyed and dangerous radiation was released into the environment. There was no evidence of health effects in the local population. Of the three examples here, it was the least serious, and yet the clean-up of the damaged reactor took about twelve years and cost nearly a billion dollars (at that period's rates), it scared the living daylights out of millions of Americans and caused major changes in safety across the industry at the time.

The Fukushima Daiichi plant in Japan was hit by a fifteen-metre high tsunami after an earthquake on the 11th of March 2011, disabling its power supply, preventing the reactors from being cooled, and three of its six reactor cores went into meltdown in the first three days. The area for twenty kilometres in every direction was evacuated and declared a no-go zone. Hundreds of employees, fire-fighters and military personnel worked heroically to contain the damage. There were 2259 deaths related to the earthquake and tsunami, but according to the World Health Organisation, no deaths were caused by the nuclear accident itself, and no-one experienced enough radiation exposure to cause sickness. However, many of those involved suffered from post-traumatic stress disorder.

As at Three Mile Island, radioactive material was released into the environment around the plant, contaminating soil and sea water. In total, 160,000 people were evacuated from their homes, and many of them had to stay away for years. While the Japanese government had lifted restrictions on most of the area by 2017, encouraging people to return, many were still reluctant to do so, and the immediate area around the plant was still off-limits, The United Nations and Greenpeace claimed the area was still showing unsafe levels of radiation. Thousands of workers are now involved in the clean-up. TEPCO, the company that owns the plant, says it will take thirty to forty years to make the site itself safe and decommission the other reactors.

Chernobyl was the worst nuclear accident in history. On the 26th of April 1986, as a result of a flawed reactor design and human error, one of the nuclear power plant's four reactors exploded. The fire burned for nine days and released a hundred times more radiation into the air than the bombs dropped on Hiroshima and Nagasaki. Some of it caught in the winds and carried across the skies over Europe.

Two people were killed in the initial blast, and twenty-eight more died soon afterwards as a result of radiation exposure while fighting the fire, with possibly another thirty in the years that followed. Hundreds of thousands of volunteers, known as 'liquidators', risked their health, even their lives, over the following years in the recovery and clean-up of the site. And 350,000 people were evacuated from a thirty-kilometre-wide exclusion zone.

Nearly thirty-five years later, there's been a lot of research into how many health problems and deaths can be traced back directly to the disaster, but there are still debates over it. So many types of illnesses could result from excessive radiation and long-term effects are very hard to judge. Some say the disaster has caused thousands of deaths, some say tens of thousands, and yet no-one can say for sure. By any measure, the accident was devastating to the land around it. The site of the explosion is now covered by a 36,000-tonne concrete and steel coffin. The clean-up of the former plant has already cost billions and it is far from over. It's estimated that the site won't be habitable for *twenty thousand years*.

Even with the contamination of the area in each case, it should be noted that the undamaged reactors at Three Mile Island continued to generate electricity until 2019. Chernobyl continued to be used after the accident too; the final reactor was only shut down fourteen years later.

One of the greatest fears of nuclear energy is not just the damage a disaster can *cause*, but how long that damage *lasts*. And even when everything's running exactly as it's supposed to, there's still the issue of the waste a nuclear plant creates. Because, while it may not contribute to global warming, the used-up fuel rods that power a nuclear reactor are fatally radioactive – and can stay that way for more than 100,000 years. If waste like this had been discovered back in the last ice age, it would still be capable of killing us *now*. We make horror films about that kind of stuff.

Short of blasting it into space, which would be mind-blowingly expensive and far too dangerous (because, y'know, the risk of exploding rockets), we have no way of disposing of it that is guaranteed to get rid of it completely. Companies that are designing underground storage facilities for this waste are having to plan further ahead than human civilisation has existed.

There's also the small matter of nuclear technology being used to create weapons of mass destruction, but let's take one global threat at a time here.

So those are the main *downsides* of nuclear power.

The alternative could be worse

Given the nightmarish problems that nuclear power presents, it might be difficult to understand why some hardcore environmentalists, people who have spent their lives campaigning to save the planet from the damage we're doing to it, are now willing to accept nuclear energy as a solution.

Which will give you some idea of just how serious the threat of climate change is to our civilisation.

Despite the very public and panic-inducing disasters involving nuclear reactors, the truth is that fossil-fuel power plants are responsible for many, many more deaths and environmental damage because of the pollution they produce. Most of us don't *notice* because it's much *slower* and it has been happening for far longer. It's normally not as visible or dramatic as an 'airpocalypse' in China, but we're still breathing the stuff in. According to a paper published by NASA, despite the three major nuclear accidents the world has experienced, nuclear power prevented an average of over 1.8 million net deaths worldwide between 1971 and 2009, because it produced massive amounts of power without all that smoke.

When we think of coal as 'cheap', the cost of the pollution isn't taken into account. If the companies that burned fossil fuels on a mass scale had to pay for all the health problems they cause in a country's population – problems that are not caused by renewables – and the damage they cause to the environment, that fuel would result in a much, *much* bigger bill. Instead, that bill is paid by the people who develop the health problems and the government that provides the health services.

The reason why nuclear power is considered worth the risk is because nothing comes close to the amount of electricity it can produce from a fraction of the amount of fuel and the resultant pollution, compared to coal, oil or gas. And for the amount of electricity you get, the fuel for a nuclear plant costs between one-third and one-fifth of a fossil-fuel plant – though nuclear comes with a whole set of other costs – and *no* greenhouse gas emissions.

The need to reduce our use of electricity is already creating a drive towards stricter regulations that are ensuring more efficient domestic appliances, better insulation in buildings and the lower bills that come with all of that. Unless we can cut back drastically on the amount of electricity we use, however, if we shut down all the fossil-fuel power plants before 2030 – which we absolutely have to do – it could take decades to get back up to the same level of power we have now just using renewables. That could mean that we would all have to find ways to use far less electricity than we do now. Nuclear power offers an alternative, for those who have it. In the space of ten years, from 1977 to 1987, France increased its nuclear power production fifteen times over, from 8 per cent to 70 per cent, with a proportional drop in the use of fossil fuels.

Some environmentalists argue that there simply is no way to achieve that kind of change just using sustainable sources of power. They claim that, given our society's appetite for electricity, the damage that climate change will do to our environment and our civilisation leaves us with no choice: we may have to include nuclear power as one of our solutions to adapting to and combating climate change.

Superpowers

The ability to produce energy on a large scale often goes hand in hand with *political and economic* power. If someone has a valuable resource, they can use their power and money to get more power and money. Most wars are fought over who controls a particular resource, whether it's water, food, land or, in recent years, reserves of oil and gas.

While nuclear reactors are cheaper to operate once they're up and running, they use incredibly complicated technology and take a massive amount of time and money to build – and then they take years to decommission afterwards, which has to be built into the cost, as does the disposal of the waste and the environmental problems it creates. Like nuclear weapons, this is just not an option for many countries. On top of this, they take *decades* to plan and build. In fact, this is another objection environmentalists have: that even if we *were* to use nuclear power to fill the gap as we close fossil-fuel plants, it would take so long for those new power stations to come online, they'd be too late to make much of a difference in the race to reduce the impacts of climate change.

The uranium for a nuclear power station has to be mined, which means similar dangers and environmental damage to other forms of mining. It then has to be refined into the form that can be used in reactors. However, compared to other forms of fuel, it's only needed in very small quantities. This means that nuclear power also cuts down on the costs and the damage done by transporting the fuel to the conventional power stations. Trucks, trains and ships are constantly moving around the world, burning diesel to carry coal, gas and oil for the power stations to burn. That's a lot of fuel being burnt, all of it adding greenhouse gases to our atmosphere.

So for those countries that already have it, nuclear technology is a *major* advantage, and it's not just being used to build power stations; it's also being used to power ships that have to spend long periods at sea. This capability extends a country's reach out further than its non-nuclear competitors. Most of these vessels are American and Russian submarines, though there are some nuclear-powered aircraft carriers too. The submarines in particular stay on patrol for months at a time, and are so self-sufficient that they only need to come back for routine maintenance, to stock up on food and to give their crews a break.

Nuclear propulsion is also used in some ice-breakers. An ice-breaker is a battering ram in the shape of a ship with a special reinforced hull, used to navigate the freezing waters and harsh conditions of the polar regions, some of them capable of ramming through ice that's over four metres thick. Russia is the only country building nuclear ice-breakers, and they are commercial, rather than military, ships. They can be used for exploration; we've learned an enormous amount about sea ice from the ships that have ploughed through it – and the submarines that have travelled underneath it, examining it with sonar. But those ice-breakers also clear paths for other ships, particularly cargo ships, through sea ice. By crossing over the top of the world, smashing *through* all that ice instead of going around it, they can make their journeys a lot shorter – and a lot cheaper. Because of the difficulties of refuelling in the Arctic, nuclear propulsion makes more sense than using a ship with diesel engines.

But Russia is also using them to serve its enormous oil industry.

In a strange twist of fate, however, ice-breakers which have been used for years to help scientific researchers explore the Arctic sea ice are becoming less necessary for crossing the Arctic Ocean as that ice thins and disappears. And this disappearing ice is also opening up the possibilities of drilling into the vast oil and gas reserves under the ocean floor – reserves that were out of reach up until now because of that ice. Russia's government makes a lot of its money from oil and gas, and wants to claim those reserves under the Arctic and take everything it can get.

In 2007, Russia dropped two submersibles to the ocean floor at the North Pole and planted a titanium flag to claim it as their territory. It was a bold move, but planting a flag in a place to say it's yours doesn't carry much legal weight any more, so nobody took it very seriously. Now, however, the country's pushing more of its military into the Arctic Ocean to support its claims. Even as the world is trying to cut back on using fossil fuels, competitors like Norway and China, as well as the United States, Canada and Denmark, will not want to surrender these resources to Russia, and will have to decide how far they're willing to go in response, either to try and claim those reserves for themselves or to stop them from being used.

The Arctic is heating up twice as fast as the rest of the world. It's ironic that the disappearance of Arctic sea ice caused by climate change, which has been caused by our use of fossil fuels, means that the world has increasing access to *even more fossil fuels.* And there could be an almighty fight over who controls them. The solution to this already exists, however, and it can be found at the other end of the planet.

At the beginning of the 20th century, Antarctica was only accessible to adventurous explorers and the most daring scientific expeditions. As more countries took an interest, some began making claims to the territory, and it looked as if there could be trouble if anyone decided to back up their claim with a military base or two. Rather than fight it out, everyone involved decided to share instead. The Antarctic Treaty was signed in Washington on the 1st of December 1959 by the twelve countries whose scientists had been active in and around Antarctica. Some of the most important points were: 1) Antarctica should only be used for peaceful purposes. 2) Freedom and co-operation had to be guaranteed for scientific investigation in Antarctica. 3) Any scientific observations and results from Antarctica should be made freely available. These conditions exist to this day.

Denmark isn't even waiting for the Arctic version of that. In 2020, the Danish parliament announced that it would ban all new oil and gas exploration in the Danish part of the North Sea and end existing production by 2050. It wants to establish the country as a front-runner in the quest for sustainable energy.

But our best hope of preventing potential conflicts over resources like oil and gas in the Arctic is to make them so irrelevant as fuels that they're not worth risking lives and huge amounts of money in the harshest environment on Earth – and ruining the Arctic's rich and delicate ecosystems in the process.

Friends of the Earth Ireland

In 2021 Ireland's electricity grid was upgraded so that 70 per cent of electricity can come from solar and wind at any one time. This is a global first! As an isolated island, managing electricity and moving it around the country as needed is a big engineering challenge.

Solar Schools are champions for renewable energy. They mostly use electricity during the day when solar panels are soaking up the sun. Many countries let you sell excess electricity back to the grid, which is a great way to make some money during holidays and weekends.

CHAPTER 16
GETTING SOMEWHERE

Every day, millions of ships and planes trace lines across the surface of the globe, joining one part of our world with another, reducing national borders to nothing but theory and paperwork. There is nowhere on the surface of our Earth that we cannot reach within a few days' travel. For a species that was travelling in dugout canoes less than ten thousand years ago, that's pretty impressive. One can only imagine what a human paddling a dugout, our very first form of transport, would think if they set eyes on a nuclear-powered submarine.

When we think of sustainable energy, we think of installations like wind turbines or rows of solar panels or perhaps one of the bigger dams, and people can raise all sorts of objections to these – that they cost too much, that they won't last, that they're too big or too ugly. People complain that it will take too much money to change over to these sources of power. Most of these installations would be dwarfed by the scale of the fossil-fuel industry's infrastructure. The difference is, most of us never see that stuff up close.

We rarely get a good look at an oil tanker or an offshore oil rig – because they're out in the ocean somewhere. And just like the wind turbines and the solar panels, governments have been helping companies like Shell, ExxonMobil and Gazprom to pay for all that gear – and if you count the health and environmental costs they've been excused from paying, they've been getting a *whole lot* more money every year than has ever been invested in renewables.

At the time of writing, the largest vessel on the ocean is the *Prelude* FLNG, owned by Shell. Technically, it's not a ship – it's a floating natural gas refinery, because though it has thrusters to manoeuvre, it had to be towed out to the position off the coast of Western Australia, where it is now moored to steel piles in the seabed. Its main deck is longer than four football fields laid end to end. Fully loaded, the structure displaces 660,000 tonnes of water, which is the equivalent of more than *five aircraft carriers*. The final cost when it's fully operational is estimated to be between $10 and $12 billion.

When people complain about the cost of investing in renewable energy, have a think about how much we are already spending on non-renewable energy. And also ask yourself: If a company has the money to build something like this, how much money can it put into spreading information it wants you to hear? How much advertising can it buy? How many articles and scientific papers can it have written? How much is it spending to control the story about energy, and about climate change? How much money can it put into supporting politicians who can change laws in its favour? Think of the power that company has.

Some of the world's largest cargo ships are pretty modest in comparison to the *Prelude*, and yet they are giants in their own right. A vessel four hundred metres long and fifty metres wide isn't unusual, over twenty storeys tall from its engine room to its bridge. These are the ships that carry most of our goods across the globe. You name the product, from rubber ducks to laptops, and more than likely, you'll find its journey to the shop involved a trip on a container ship. These massive vessels are the most economical way of transporting anything in bulk, and the world relies on them to deliver the goods.

Because they spend most of their time out of sight of land, however, they can get away with a lot more pollution than we'd tolerate from trucks, trains or cars. Shipping is responsible for 2.5 to 3 per cent of the world's air pollution. Three per cent doesn't sound like a lot. To put it in perspective, one tonne of carbon dioxide stored at normal atmospheric pressure would fill a small house. Three per cent of the world's emissions comes to about *940 million tonnes* of CO_2.

Up until recently, container ships ran on a pitch-black treacle known as 'bunker fuel', the dirtiest, most polluting form of diesel there is. The annual emissions of one large container ship was comparable to the emissions from ten to fifteen million cars running on diesel. A new international law brought in in 2020 is now forcing these ships to use cleaner fuel, but that is aimed at reducing sulphur pollution, and won't do much to cut greenhouse gases. And yet, there is a will to make a change. In 2018, Maersk, the world's largest container shipping company, committed to making its operations carbon neutral by 2050, by whatever means possible. If they manage it, their competitors may be forced to follow their example.

While shipping burns fuel on an awesome scale, the aviation industry gives it a run for its money, providing over 2 per cent of the world's air pollution, which is equivalent to the emissions of a major country. Airplanes can't carry anything like the numbers of people or cargo that ships can, but they can carry them much, *much* faster. It's why people prefer to fly rather than sail – and do so in their billions every year. Most cargo, on the other hand, isn't in any major rush, and is happy to take the boat.

Fuel is the biggest cost for anyone who runs a jet airliner, so these companies are constantly looking for ways to burn less of the stuff, and some have even begun testing biofuels, though this is still at the very early stages. The Boeing 747 is an old plane, but can still carry more passengers than many younger airliners. It needs nearly 8000kg of jet fuel to take off and land, which is ... a *lot*. Once it's cruising, however, it needs just 10kg for every extra kilometre it flies – and it can carry more than four hundred people. For a distance of a few hundred kilometres, it would cost and pollute far less just to drive those people in cars, and even less in buses or trains. It's only once you get into the *thousands* of kilometres that flying starts to make more sense in terms of cost and pollution.

The question we need to ask then is: Who really *needs* to fly thousands of kilometres, why, and how often? And would they really be willing to travel that distance if they had to take another, much slower form of transport?

If there's one thing that the COVID pandemic has taught us, it's that a lot of business we used to travel for can now be done in the office or even at home. The more communications technology develops, the more it reduces our need for transport – and the emissions it produces.

Please place your seat backs and tray tables in their fully upright position. It's time to aim higher.

Precious metals

About 25 per cent of all global emissions of carbon dioxide comes from transport, and *three-quarters* of that chunk is given off by road transport. The only form of transport where emissions have dropped in the last few years (not including the drop caused by the pandemic) is railways, because many trains are now being driven by electricity rather than diesel – though, of course, that electricity has to come from the supply network, which still mostly means power plants.

Europe is made up of some of the wealthiest countries in the world, and we have become very used to getting around with speed, convenience and personal space. We *love* our cars. Unfortunately, they're a bit *too* convenient, and we often use them for just one or two people. I am as guilty of this as anyone; I live in the countryside where there is very little public transport and most of my travelling is done by car. If you leave out over-the-top things like personal jets or helicopters – which the vast majority of us don't really have access to – a car with one person in it is the most polluting form of transport there is for the distance a single person travels.

Electric and hybrid cars (which have both conventional and electric motors) are becoming more popular and more affordable. Fully electric cars are very much part of the solution; they do not emit any greenhouse gases themselves and they cost much less than a normal car to run, but the number of charging points, even in a relatively wealthy country like Ireland, isn't anywhere close to the networks of fossil-fuel service stations, and for many people at the moment, buying an electric car is still an expensive option.

Hybrids still burn fuel – though much less than conventional cars, if they're used properly – and electric cars are charged from an electricity network that is still mostly powered by fossil fuels, so it's as important to change the infrastructure around them as it is to change the cars themselves.

Along with building all those charging points from scratch, another option is to convert existing service stations to fuel hydrogen-powered vehicles. The waste product from a hydrogen-fuelled motor is … wait for it … *water vapour*. That's how clean it is. There are already cars and boats that run on hydrogen. At the moment, however, most of our hydrogen is created from fossil fuels – particularly natural gas. On the other hand, there's a much more expensive method that can make it from water by a process called electrolysis, and this can be powered by renewable energy, a technology that is developing quickly. Hydrogen takes up more space as a fuel than oil, requiring larger tanks; it has to be stored as a super-cooled liquid or a highly compressed gas, and is incredibly flammable (check out the story of the *Hindenburg*). But it's seen as having huge potential, because battery-driven motors aren't the solution to everything, so it could be a more practical alternative for aircraft and large ships.

While electric motors are a big improvement on petrol or diesel ones, they don't yet offer much of an alternative for aircraft. Batteries are *heavy*, and don't produce nearly as much power for their weight as jet fuel, which is a bit of an issue for flying, where every gram of weight matters. We're a long way from having practical electric aircraft, and nearly as far from having batteries with enough muscle to power ships for long periods, though these are already in development.

Batteries also come with one other problem: how they are made.

The batteries in electric vehicles are pretty complex, using metals such as cobalt, lithium, nickel and others. In a world that is demanding batteries for more and more forms of electronics, the mines that produce these materials are struggling to cope with demand. They just can't dig the stuff out of the ground fast enough. For instance, the Irish government has set a target of having 950,000 electric vehicles on Irish roads by 2030. This is quite the ambition, and would take a nice big bite out of all the carbon dioxide we're emitting, but there are major questions about how realistic it is. In 2018, there were fewer than five thousand electric cars in Ireland. With the restricted supply in batteries, many in the car industry believe that it will be impossible to produce and sell that many electric cars in that time-scale. They simply can't make them that fast.

And that's just for Ireland, which is a very small country that has to compete for those cars with much bigger, wealthier markets. Every petrol or diesel car off the road is a good thing, it means a little less greenhouse gas in the air, but replacing all of them with electric cars just won't be possible – and that's before you look at the moral and environmental issues around how these batteries are made. Because relying on these materials means damage to the environment by some of the most destructive forms of mining, and damage to people working in utterly horrible conditions.

Let's take cobalt as just one example. Half of the cobalt in the world comes from the Democratic Republic of Congo. Many of these mines are worked by hand, without heavy machinery, and some of this work is carried out using child labour. Children as young as seven are sent down holes in the ground for as little as one dollar a day to dig for the substance used to make our phones, laptops, tablets and car batteries. Our money is supporting this abuse of children. We have to plan a future based on batteries which ensures just treatment for the communities who provide the materials for them. For obvious reasons, car batteries need a *lot* more cobalt than a phone battery, so researchers are already looking for synthetic alternatives. Panasonic, who make the batteries for Tesla cars, announced in 2018 that they were developing batteries that wouldn't need cobalt.

We need to share more

There's no question that we need electric vehicles to tackle global warming, but we also have to make sure that other people aren't paying the worst of the cost on our behalf. This will become an increasingly important issue because the more we rely on sustainable energy like wind and solar, which don't provide a steady supply all the time, the more we'll need batteries to store that energy and release it as we need it. It's vital that our appetite for those precious materials doesn't make conditions worse for the environment and for the people who work in and live around those mines.

If we cannot replace all the vehicles we have now with electric vehicles, then we have to cut down on our use of vehicles. We can do less damage by slowing down the pace of life, accepting a bit of inconvenience and sharing resources more, such as using much more public transport. You don't have to talk to anyone on the bus or train if you don't want to – that's what books, phones and headphones are for. Every nation in the world needs to make public transport a priority.

A bus carrying fifty people can mean *forty-nine cars* are not being used for that journey. Now expand that out to a train carrying *two hundred* people. But those services *have to be there* for us to use them. We also need to ask ourselves what journeys we have to take at all, or if we need a vehicle to do it. Can we walk or cycle that short trip to the shops, instead of taking the car? Is it safe for children to cycle to school? Are bikes available to hire, if someone wants to cycle around town? In Ireland and Britain, though cycle lanes are starting to be given priority in cities, we are still well behind countries like the Netherlands and Denmark, where walking and cycling are encouraged far more, and they have the infrastructure to back it up. These are simpler, cheaper, more practical and healthier solutions than constantly trying to adapt our towns and cities to increasing numbers of cars. Do we have to fly somewhere for a holiday, or could we take a boat, or could we holiday closer to home? Do we really need to make that business trip, or could that meeting be held online?

The COVID pandemic has given us a preview of this. In 2020, carbon dioxide emissions dropped by 7 per cent, mainly because people around the world did much less travelling. The UN says *we have to achieve that kind of drop every year until 2030* to keep the rise in global temperatures to 1.5° Celsius. So essentially, we have to take action, not by doing more, but by doing *less*.

The cost of using transport is not just the fuel, or the ticket, or the maintenance, or the price of the vehicle, or the other figures we associate with owning or travelling in a vehicle. If we were all forced to pay for the health and environmental damage caused by the pollution from transport, we'd put a lot more thought into how we travelled. Instead, thanks to our changing climate, someone somewhere else could be paying the unseen costs of our modern lifestyle.

It is very hard for an individual to make a difference in this situation, but when large groups of individuals get together, their money and choices begin to have an effect – from a family to a community, to a village, town, city or country. The more people who make a decision not to buy a particular product or use a particular service – or the more they demand a new one – the more likely they are to be listened to. Whether we're dealing with a government or a multinational company, the bigger and louder the crowd, the greater the power.

But of course, it's not just people that travel – it's pretty much all the things we use in our lives too, getting from where they are made to where we can buy them. And a lot of that cargo is just a complete waste of space, fuel and money – especially considering where it goes after we're finished with it.

Many schools have 'walk or cycle to school' groups to encourage exercise and reduce car pollution. Plan a low impact family staycation like camping or hiking in the mountains instead of going abroad to reduce your travel footprint. Sail-and-rail is also a great option instead of flying, and it's fun!

CHAPTER 17
WASTED

Humans produce a lot of waste. I don't mean poo – though we do produce a lot of *that* too – I mean the crap that's left over after we've finished doing whatever we're doing. We're buying more and more stuff every year, and the more things we buy, the more we use and the more we throw away. As populations grow and increasing numbers of us live in cities, the amount of waste we generate is set to jump from 2.01 billion tonnes in 2016 to 3.4 billion tonnes over the next thirty years.

That's a lot of stuff to get rid of.

In high-income countries, more than one-third of waste is recovered through recycling and composting; only 4 per cent of waste in low-income countries is recycled. Try and imagine what has to be done with all that rubbish. It has to *go* somewhere. Most of it gets dumped and some of it is burned. If it gets dumped, it's using up huge areas of land and giving off gases as it rots and breaks down. Only human civilisation could create vast landscapes of garbage, stretching to the horizon, poisoning the air and the ground. If that garbage is burned, it's releasing even more fumes into the air, though at least we can use that kind of giant dumpster fire to generate electricity. Most of us generally don't spend much time worrying about these things, because most of us don't live next to a landfill or an incinerator. It's someone else's problem.

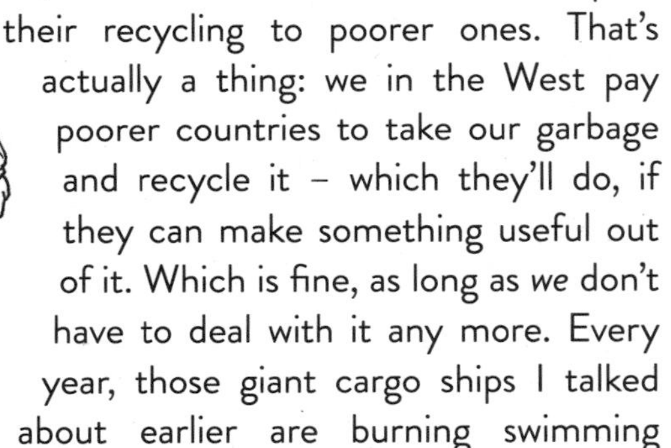

In fact, richer countries can and do export their recycling to poorer ones. That's actually a thing: we in the West pay poorer countries to take our garbage and recycle it – which they'll do, if they can make something useful out of it. Which is fine, as long as *we* don't have to deal with it any more. Every year, those giant cargo ships I talked about earlier are burning swimming pools of dirty diesel to carry rubbish from one country to another. It's moving your mess from one room to another and thinking it's sorted – but on a global scale.

But there is nothing separating the air that that garbage is poisoning, and the air we're all breathing. It's estimated that the equivalent of 1.6 billion tonnes of carbon dioxide was generated from the treatment and disposal of waste in 2016 – which is about 5 per cent of global emissions. Nearly all the emissions from shipping and aviation put together. Our love for plastic, however, causes an extra special problem.

The miracle material

Plastic is an absolute wonder. It can be formed into almost any shape, can be made in any colour, it's flexible, can be watertight and airtight and is used in everything from toothbrushes to telescopes, from disposable packaging to components in the International Space Station. It reduces food waste by keeping it fresh. It can save lives by keeping medical products sterile. It's under my fingertips as I type this.

Given that it's such a fantastic material, you'd think we'd value it more. And yet, in 2016, 12 per cent of all solid waste was plastic – about 242 million tonnes of the stuff. We just *threw it out*. It's made from fossil fuels that can't be replaced, and we're tossing most of it into landfills or incinerators or the sea. We recycle some of it, but even then, it's degrading every time. A plastic bottle can't be made into another plastic bottle; it's more likely to be used in the manufacture of fabric for clothing.

One of the best qualities of plastic is also what makes it a nightmare for the environment. It's durable. It can last a *really* long time. Some types of plastic take hundreds of years to break down, while others break down into tiny particles that then stay in the ecosystem and end up becoming concentrated in the food chain, as small animals eat plants and bigger animals eat smaller animals. Which is how we are ending up with tiny particles of plastic in the food we're eating every day. Though it does leach toxic chemicals as it degrades, in terms of climate change, plastic waste is not a big contributor to greenhouse gases – though it can be if it's incinerated – and the convenience of plastic did lead to our using less glass. Glass is not made from fossil fuels, we tend to reuse it more often, and it is completely recyclable.

By making so much of the stuff we need out of plastic, we are taking a material made from a resource *we can't replace*, and we're using it as if it will never run out. After the emissions given off by the oil and gas industries in making the raw material, we are then burning fuel to generate energy for factories that shape this limited resource into things we'll often only use once. We're burning *more* fuel to transport it around the world, and once we're done with it, we throw it away, with little thought of where it goes and what happens to it after that. In many cases, we'll burn *even more fuel* to transport it away from where it was thrown out, so that it can be dumped into the environment, where it will cause damage for a few hundred years – or perhaps we'll end up swallowing it with our food.

Many of the things we make from plastic are used once and thrown away. And it's not just plastic; it's estimated that, of the materials flowing through the economy, only 1 per cent are still in use six months after they were bought. Just 1 per cent. Even the goods we might have expected to hold onto are thrown away because they've been damaged, or they're just unfashionable.

This is a problem we can all contribute to solving. *We can just buy less stuff.* It really is that simple. It would even save us money.

Some day, archaeologists and even geologists may look back at our civilisation, at the traces of it in the soil and rock, and will identify it by the astonishing layer of waste we've left across the surface of our planet. Our activity has had such a lasting effect that there is now a call to name this geological epoch the Anthropocene, from 'anthropos', the ancient Greek word for 'human'. The current epoch is the Holocene, but there are many in the scientific community who argue that the start of regular nuclear weapons testing in the 1950s marks what should be the beginning of the Anthropocene, a new chapter in the Earth's development.

The building blocks of civilisation

Another material that may endure long after we're gone is concrete. It is the most widely used man-made material in existence. And its main component, cement, is having a major impact on our environment. Everywhere you look in a town or city, you'll see buildings made using concrete, or using bricks held together with cement. The manufacture of cement has a serious carbon footprint – as much as 8 per cent of our total global emissions. In 2016, it came to about 2.2 billion tonnes of CO_2 – more than aviation and shipping combined. Made mainly from limestone and clay, it's so vital for the construction industry, we rarely consider it as something we can cut back on. There are alternative ways of building, though few of them are as affordable, efficient, versatile and durable.

The search is already on to find building materials to replace cement and concrete. There is a type of cement already used in construction that produces far less carbon dioxide and can be used on its own or mixed with the normal stuff (Portland cement). This new kind of cement is known

as 'ground granulated blast furnace slag' – or GGBFS to its friends – and it's whiter and potentially stronger than Portland cement. And it's made by recycling waste produced by the iron industry.

Needless to say, the more a country is building and manufacturing, the more it's transporting, the more its people are travelling, the more it's burning and the more waste it's throwing away. And therefore, the more responsibility that society has to make things right.

But that society may not be made to suffer from the consequences of climate change as much as others who have contributed far less to the problem.

Break Free from Plastic is a global movement of communities taking action on plastic. Saying no to plastic is one of the easiest ways to have a positive impact – start by refusing plastic in supermarkets and asking local businesses to provide alternatives like offering loose produce and reusable containers.

CHAPTER 18
HURT FIRST, HURT WORST

It is a bitter irony that many of the countries that are already being hit by the worst of what climate change has to offer are the countries that have contributed the least to causing it. They may well have been colonised in the past and had their resources stolen for generations – with effects that are still being felt to this day – slowing their development even further. The richest nations in the world got to that position by burning their way to a more industrialised society, with many of them having lots of free labour, in the form of slaves. From the Industrial Revolution onwards, development required the generation of energy and that demanded fossil fuels. The average American, now living in a country where wealth and the consumer lifestyle is at its peak, releases 16.2 tonnes of carbon dioxide per year. Compare that with people in some of the poorest countries in sub-Saharan Africa – such as Chad, Niger and the Central African Republic – where the average carbon footprint per person is around 0.1 tonnes per year. That's about what an American emits in two and a half days.

On the other hand, wealth also means the ability to make big changes when you need to. In France, the average person has a footprint of 5.5. tonnes. In the UK, it's 5.8 tonnes. They were able to lower their emissions because they had nuclear power, and the resources to shift their power sources towards renewables, and they continue to move in that direction. Wealth gives you options. Many of the poorest countries don't have these resources – and continue to lose what they have to richer countries. They are struggling to adapt, even as they endure costly disasters caused by dramatic changes in their weather.

Whether it's the hurricanes that rage across Puerto Rico and Dominica, or Sri Lanka facing heavy landslides and flooding, or severe drought in Yemen, countries that did little to create this disaster are facing its worst effects. Like our garbage, the industrialised world has exported the most dire consequences of our carbon addiction to our less prosperous neighbours.

Imagine being forced from your home

Bangladesh has a population of 165 million, and one-third of them live along the fertile southern coast; and their economy centres on farming and fishing. Most of the land is barely above sea level. During the rainy season, flood waters can cover one-fifth of the country. Now the unpredictable swings between drought and heavy rain, combined with rising sea levels, are causing serious erosion, destroying crops, washing away soil and contaminating land with salt. The sea is coming for

them. It's estimated that as many as 13.3 million people could have to leave their land by 2050.

That's just one country.

Countries on or near the equator face summers with merciless heat. The World Bank estimates that there are 143 million people across South Asia, sub-Saharan Africa and Latin America who are at risk of being driven from their homes as conditions become impossible to live in. And those people are going to have to go somewhere.

Leaving your home for a holiday takes more money and more effort than your normal life. It's the kind of thing most of us have to save up for. Now, imagine conditions have been tough for a long time, so you're already close to broke. Perhaps it's so bad that war has broken out, and you fear for your life, and the lives of your family. And now you have to leave *forever*. You may have a car, but you'll probably have to abandon it eventually, so of all the things you own, you can only bring what you can carry. What would you choose to bring? Where will you sleep? How do you even do something as basic as keep your phone charged?

You have to travel until you can find somewhere else to live, but there are so many like you, all facing the same problem, and you have so little money, you may soon be starving or relying on the charity of others to survive.

The more desperate you become, the more vulnerable you are to criminals who will take advantage of you. Time and again, you will arrive in a new town or city and be told there is no place for you there. You could end up in a crowded refugee camp or you may be able to keep travelling, but sooner or later, you will run out of money, or you'll reach a place you can't get out of, or you'll run into people who stop you from going any further. They may be violent towards you, when all you're trying to do is stay alive. What would you be willing to do to stay alive, or to save the ones you love?

When we talk about 'climate justice', we mean this threat to the vulnerable and what we can do to reduce the harm being done to them. There are millions of people already facing situations like this, and as the climate crisis worsens, there will be even more. Any plan to reduce the damage climate change is doing or to help the world adapt to the changes we face has to include support for people struggling to live normal lives in some of the hardest hit regions of the world. And it must include humane and practical solutions for those who have been forced to leave their homes.

This is a whole-world problem, and needs to be tackled on a whole-world basis, with reason, compassion and justice at its heart.

Friends of the Earth Ireland

Fast Fashion is bad for people and nature – clothes are made through cheap labour and the process creates huge environmental pollution. Slow Fashion is a better option – buy second hand or share clothes with friends. Buy less, and check for brands that care for people and nature. Have a browse at: www.fashionrevolution.org

CHAPTER 19
TIPPING POINTS

This is the most negative chapter of this book. If you find that thinking about this stuff makes you anxious, you can skip this next bit if you want – the book works fine without it. I say that because this next section is about the things that give climate scientists nightmares.

Climate change is normally described as a gradual process that happens over decades, but there are certain situations where that slow change can push conditions to the point where a sudden major event causes things to speed up, creating a domino effect, meaning changes we expected over a century or two could happen over a matter of years. These are known as 'tipping points' or sometimes as 'feedback loops'. They are also a good illustration of how small changes, multiplied many times over, can result in extraordinary events. Remember what I said in the introduction: humans did this, and strange though it seems, there is hope to be found in that fact. The changes that created these situations can be traced back to the unstoppable growth and appetites of our civilisation. And for the vast majority of that time, we were less knowledgeable, less enlightened and capable of far less than we are now.

Now the species that changed the planet's climate is awakening to the problem, and is slowly but steadily turning its considerable collective intelligence and power to dealing with it. Billions of small acts can have major effects, both good and bad. The relationships between all the things affecting the climate are such a tangled mess that it would boggle the brain, but I'll cover some of the best-known tipping points here.

The disappearance of Arctic Sea ice

I've mentioned this in other sections of the book, but it's a biggie, so I'm bringing it up again here. White ice reflects sunlight; dark ocean absorbs it. That's the albedo effect. Rising temperatures mean less ice, which means higher temperatures, which means less ice, over and over until there's no ice at all, and it doesn't come back. Once it's gone, that dark ocean will just keep absorbing more heat, speeding everything up.

Melting permafrost

Permafrost refers to ground that is permanently frozen, like much of the land in the Arctic Circle. That land is now starting to thaw out. An early sign of this is 'drunken trees', where the frozen ground has softened so much that trees start leaning over. The soil under the boreal forests in these regions is thought to be one of the largest stores of carbon in the world. That land is also saturated with methane, held in place by the ice. Methane is thirty times more potent at warming the planet than carbon dioxide over a hundred-year period, so the more methane that is released, the more warming there will be, which will melt more permafrost, which will release more methane, and on and on in another vicious cycle.

The Amazon

Rises in temperature are also having the effect of causing droughts and increasing numbers of wildfires in the Amazon rainforest. The trees are drying out, even as large areas of the forest are being cut down for wood and to create more farmland. The forest is so big, it affects the weather, cycling moisture down into the ground and back out into the air again. With less forest, there is less rain. Less rain means less vegetation means less rain means dryer vegetation, which is more likely to catch fire, which means less vegetation, which means less rain. You can see how this might be a problem, given that we need trees to take in carbon dioxide and release oxygen.

Coral reef die-offs

This is something else that was covered earlier, but it's worth mentioning again. As sea temperatures rise, it kills off

the algae that feed the coral, and then the coral die off, and all the small creatures that live on the reef die off or leave, and this carries on up the food chain. One small change in temperature wipes out an entire ecosystem, as well as the livelihoods of coastal communities that rely on the reefs for fishing and tourism. Ocean plants take in carbon dioxide and give off oxygen, just like land plants. As the oceans grow hotter and more acidic, life in the water dies off, and it affects our carbon cycle in a major way – which makes conditions in the oceans worse.

Melting glaciers

As glaciers melt, they create the same problem as the disappearing sea ice: there's less bright ice to reflect sunlight, and more dark ground to absorb it. The breakdown of the glaciers, particularly in Greenland and the Antarctic, is now happening faster than scientists predicted. Over and over again, we are seeing images of cliff faces of ice collapsing into the sea. This is contributing to rising sea levels, as billions of tonnes of ice empty into the oceans every year, but because it's fresh water, it could also lead to another humdinger of a problem ...

The slowdown of the thermohaline circulation

The great ocean conveyor belt – that slow, massive current that helps distribute heat around the world's oceans – is driven by contrasts between hot and cold, and between fresh and salty water. Water freezing into ice as it flows up through the polar regions makes the water around it saltier. Salt water is denser and sinks, pulling in the water behind it. If there is less freezing, then you get less of that concentration of salt in the water, and less *movement* of water. This situation is made worse if nearby glaciers are emptying fresh water into the sea, reducing that concentration of salt even further. If there is less contrast between warm and cold regions, this also slows down the current.

The conveyor belt actually stalled back in the ice age, though it's thought that's unlikely to happen again, at least not in the near future. But a slower flow means the water isn't mixing as much – so it's not pulling heat from the surface down to the depths, and therefore the oceans won't soak up as much heat as before. This is a problem, because so far, the oceans have absorbed most of the extra heat that climate change has caused. The more this slowdown of the cycle happens, the more it will keep happening.

There are two really important points to make about all of these situations. The first is that there is no way that one group of people, or one country or continent, can stop or reduce the damage from these disasters on their own. Every nation in the world has to get involved.

The second point is that these situations are so big and so slow to develop that most of us just can't see them coming. They are also vast subjects, covering many areas of science, and none of us can know everything. We have to listen to the experts who study these things, who've spent their lives learning how they work. That's what experts are for.

Friends of the Earth Ireland

This could be a good time to go for a walk outside. It's good to remind ourselves how amazing nature is. If you feel worried about climate change you can talk to a friend. Chances are you're not alone. We feel sad because we care – and that's a good thing.

CHAPTER 20
WHY DO PEOPLE DENY THE SCIENCE?

Sometimes you'll hear a debate between someone talking about the climate crisis and someone else saying there's no such thing, as if there's a fifty-fifty split in opinion among experts. This is simply not the case. The science confirming the climate crisis and its effects is overwhelming. It has been confirmed by millions of scientists in every area of research concerned with the Earth and its environment. Any argument about whether it is really happening is well and truly over. So why are there still so many people who don't 'believe' in climate change, or who don't think they need to act on it?

The types of scepticism range from the average person who just hasn't grasped the seriousness of the situation, right up to the self-serving, corrupt billionaire who knows but doesn't care, as long as their profits aren't affected. It doesn't help that scientists are *cautious* about the language they use, saying things like: 'Looking at the available evidence, there's a high chance of a three-degree rise in global temperatures by the end of the century.' It makes it sound like they're *not very sure*, but this is because science has extremely high standards about things like facts. Science is so rooted in questioning and investigation that very little is stated in absolute terms, which can be misinterpreted as doubt. As well as this, they are often taking a very detailed process of measuring something very complicated and trying to boil it down into simple figures most people can understand.

On the other hand, it's a lot easier for someone who just wants to cherry-pick the single fact that suits their argument, like: 'This winter was colder than last winter, so the world's not warming up.' And if you're just plain *lying* about it, well ... then you can sound as certain as you like, because being accurate doesn't matter, and explaining why it's a lie will often take more time than guests have in a short television or radio interview. The fact-checker is always left chasing after the lie.

Let's look at the most common forms of denial, and what motivates them.

'Not my problem'

For a large proportion of people, the issue of global warming feels too theoretical, too distant, for them to feel any connection to it. It's only over the last ten or fifteen years that it has become a subject in day-to-day conversation for people who aren't specifically interested in it – as part of our normality. You'll remember normality bias from earlier in this book. It plays a big part in this. We

assume our normality won't change. But there's also the problem that our brains – which haven't evolved in any major way since our stone age days – are wired to respond to short-term, immediate threats, rather than drawn-out, distant ones.

For a long time, a lot of the news that people heard about climate change involved places they'd never been to, like Antarctica or Greenland or the upper atmosphere. And even though we're now seeing the effects on real people in real time, it still involves complicated scientific data that few people outside of those areas of research understand. There's only so much we can worry about, and it's hard to feel any urgency about something so long-term. It just doesn't get us on an emotional level – so it doesn't *motivate* us. If someone *does* grasp how big a threat this is, they may well feel overwhelmed by it, and choose to just ignore it, hoping it will go away, because they don't know how to deal with it. Feeling disconnected from it prevents them from changing anything. They might also feel that it's something that isn't going to happen for a long time, and that it is a problem for politicians and scientists to solve, rather than, y'know ... *everyone*.

'I don't want to go against everyone else'

The things we consider important are often defined by what is important to our immediate family, social group and community, and on a wider basis, by those we identify with, whether that's through politics, religion, nationality or some other bond. Some of this is conscious – we know it's having an effect – while there'll be other times we weren't even aware of it until we mix with, or just have conversations with, people from different backgrounds.

While you might think science should be above all this, as it's supposed to be based on facts, we are emotional beings more than rational ones, and in the same way we're better at reacting to close-in, immediate threats, we are also more likely to be persuaded by someone we know and respect than we are by someone we don't know, but who is an expert in their particular field. And if we're part of a social group where belief and loyalty to the group is valued over reason – and there are many different kinds of these groups – then anything that science says that *challenges* those beliefs could be rejected.

If someone believes, for instance, in the conspiracy theory that the COVID-19 virus is a worldwide hoax aimed at making them take a vaccine that will be used to control their minds, they can come to see themselves as part of a resistance against a conspiracy run by powerful people. Identifying themselves as the resistance is an inherent part of who they are. They are fighting against their oppressors. They will surround themselves with others who feel the same way, and they all bond on that basis, reinforcing each other's beliefs.

So now it is not just an argument about the science; contradicting them can seem that you're questioning this person's very *identity* – who they believe themselves to be. And not just them, but many of the people they love most in the world. It makes them defensive and closed off. The more threatening the information is to their beliefs, the less they listen.

And once these arguments become a part of how the whole group identifies itself – 'we believe we are the resistance fighting a conspiracy to oppress us and control our minds with this vaccine' – then just trying to show them evidence that they're wrong makes you an enemy, and it's hard to have any reasonable discussion on that subject at all. Once people start down that road, *any* information, no matter how inaccurate or downright fake, that supports the beliefs of the group becomes more important than facts. While this has gone on for as long as there have been people, with social media, this false information no longer just travels by word of mouth – now it's all over the internet.

This is reinforced by a phenomenon known as 'confirmation bias', where you're more likely to believe some new information if it confirms what you already think – something we're *all* prone to doing. But once you've undermined the very idea of rational discussion, trying to have an argument based on the science has little effect.

'I just don't care about the rest of you'

Tobacco companies knew that cigarettes killed people as early as 1953, but they covered that information up, and lied about the effects of smoking, so that they could keep making huge profits from selling cigarettes. The fossil-fuel companies, who spend a fortune researching their own product, knew about the threat of climate change in the 1970s. Journalists found a memo from 1979 that was sent around to ExxonMobil's leading scientists, warning that the use of fossil fuels would have major effects on our climate. This was not some raving environmentalist shouting at the gates; this came from the *researchers in an oil company*. In 1979. In fact, there's evidence they could have known as early as 1957. And not only did they keep a lid on it, they spent millions of dollars covering it up and spreading disinformation, until they were finally exposed. And these companies still give eye-watering sums of money to politicians and PR people around the world to hold back efforts to tackle climate change.

There are many people who refuse to accept the evidence of climate change because they can't get their heads around it. There are some who refuse to accept it because they feel it conflicts with their beliefs and the values of their social group.

But there are those who accept the evidence, and *they just don't care*, because of their appetite for money and power. People who feel no responsibility to the world around them. All that matters to them is their own needs, and the needs of those closest to them. They are wealthy enough that they can insulate themselves from the worst of the effects – at least for the moment – and have no intention of helping anyone else, or doing anything that doesn't benefit them directly. And unfortunately, some of the richest and most powerful people in the world fall into this category. And they understand the power of storytelling, of communication. They spend colossal amounts of money in the media trying to deny and confuse the science. And that money can make their voices louder than their opponents'. Their influence can include control of news organisations, research funding and even politicians.

But these people *do not* control the world, and the COVID-19 pandemic has shown us absolute proof that they cannot isolate themselves from it. A virus doesn't care how much you have in your bank account. The lives of the world's most powerful people are built on the backs of others, and they are only as secure as the people who support their privileged existence. We all experience the same weather, our food all comes from the same land and sea, and we all breathe the same air. You can't live in a different world to the person who serves you your food or collects your garbage. Like the coronavirus, no matter who you are, sooner or later, our changing climate will come for you too.

Friends of the Earth Ireland

We sometimes forget that the government works for us. If we want things to change, we need to tell them. And the more people that speak up, the better response we will get. You can write to your local politician and tell them about the change you'd like to see.

People around the world are taking their governments to court for not protecting the climate. And cases taken against the government in the Netherlands and Ireland have been won! In 2020 there were 1400 such legal battles ongoing in thirty-three countries.

CHAPTER 21
THIS IS A WAR WE WIN BY DOING GOOD

Fire gave us our civilisation and allowed us to become the advanced species that we are now, but that couldn't go on for ever. When you get right down to it, fire *burns* fuel faster than we can *make* fuel, so any society based on burning stuff has a limited future – and now we're starting to see the end of the line. If it was freezing cold outside, you could burn all the furniture in your house to keep yourself warm, but you'd quickly run out of fuel, and then you'd be cold and have no furniture. We are taking more than the Earth can afford to give.

There have been times in every nation's history when it had to mobilise for war, where the entire population had to work towards a single goal. To prevent the worst consequences of climate change, we need the nations of the world to mobilise in the same way. Unlike any war that has ever been fought before, however, this one will save lives rather than destroy them. Our environment will heal and grow, rather than suffer burnt scars, rubble and ruin.

The feel-good factor

There's a whole load of negative stuff in this book, and yet there is a lot to feel good about too. There are many reasons to believe that not only can we pull our civilisation back from the cliff edge, but we can start living in a way that does good for our planet, rather than doing harm. It's important to realise that, because we are causing the problem, we have the power to fix it. It's not a world-disrupting solar flare or a super-volcano or an incoming asteroid. It's *us*. And if *we* are doing it, then we can *stop* doing it.

The movement to meet the challenge of climate change is the biggest the world has ever seen. People in every country around the globe are taking action; millions are undertaking a multitude of small tasks contributing to this single purpose. There has never been an issue that mobilised Earth's population on this scale, and it is a fantastically empowering feeling to be part of that.

There is no better demonstration of this than the School Strike for Climate. This movement was largely inspired by the Swedish teenager Greta Thunberg. In 2018, 15-year-old Greta started a school strike for climate action in Sweden. Since then students around the world have joined her. In March 2019, the first global strike was held, with thousands of school strikes in over a hundred countries. More events followed, with millions of people taking part in protest marches, many of them schoolchildren. The event in September 2019 is thought to have been the largest climate protest in world history, with as many as six million taking part, and represents a generation of youth activists that have energised the fight for our environment.

Greta and the school strikers have one simple message for political leaders: Stop the climate and ecological crisis. Speaking to the World Economic Forum in Davos in 2021, Greta said: 'Right now more than ever we are desperate for hope. But what is hope? For me, hope is the feeling that keeps you going, even though all odds may be against you. For me hope comes from action not just words. For me, hope is telling it like it is.'

Most of the solutions that will help us face the challenge of climate change are good for the human race too. They are things we should be doing anyway. Finding sources of energy that don't come from fossil fuels means finding energy generation that doesn't pollute the air we breathe and will never run out. Finding ways of cutting down on our electricity means smaller bills. Finding ways of farming that work in harmony with the environment, rather than fighting against it, makes the whole process more sustainable. It helps ensure our food supply is secure and makes farmers less reliant on subsidies from governments and on major corporations for seeds, fertilisers, pesticides and fungicides.

Shopping local is good for the community and will reduce the amount of products being transported around the world. This will also be helped by cutting back on our consumerist culture, fixing things rather than buying new ones, which means spending less money, placing more value on what we *do* have, and creating less waste.

Even human rights are bound up in the way we treat our environment. Y'know, that whole thing about treating people with basic decency. We are making that better too. It's all woven together.

Start small and find your own thing

Looking at the big picture, this can all feel overwhelming, because for most of us 'big picture' is not how we do things. We need to deal with problems in small, manageable chunks – so, that's where we start. Ask yourself, what are *you* interested in? Don't just take my word for anything in this book. Pick something you have a passion for and find out more for yourself. And it's important to realise that we are not just starting *now*. The changes we need to make have been happening for decades.

There's a lot you can do on a day-to-day basis, following the principles of 'reduce, reuse, recycle'. Use less. Fix things rather than buying new ones. Resist the 'use-once-and-throw-away' culture. Recycle more. You can't *know* everything and you can't *do* everything, but we can all do *something*. And it'll work best if it's something you're already interested in.

What you can't do on your own, you might be able to achieve with a group of people. Get involved with things. Many practical improvements are being made at a local community level, where that community sees the immediate benefits.

Here are some examples.

- The Eco-Schools programme teaches young people about environmental issues and gets them involved in projects where they put that knowledge into practice. It's run in Ireland as Green-Schools, by An Taisce. The programme starts in the classroom and expands to the school, with the aim of inspiring change in the community at large. Eco-Schools is currently operating in sixty-two countries worldwide with over sixteen million students taking part. It's a great introduction to the possibilities, and a way for young people to find out what areas interest them, and what they might like to focus on. It's also a good example of information being shared around the globe, people from different countries sharing resources and helping each other solve problems.
- In 2012, out of the 915 million people living in sub-Saharan Africa, about 730 million of them used traditional fuel such as wood and dung for cooking purposes. Breathing in all that smoke has some

pretty bad effects on their health, but for many, they don't have a choice, because they don't have access to a reliable electricity network. In some of these countries, less than 50 per cent of the population have access. In Sierra Leone, more than three-quarters of the population don't have a reliable supply. This also means no refrigeration for food or vital medicines, and the only light for activities at night comes from burning oil lamps or candles, which means more smoke and a greater risk of fire.

Local programmes are now introducing solar technology in these communities, providing that electricity, as well as creating new jobs for young people. Water supplies that once relied on diesel-powered pumps to pull water up from wells can now use solar-powered ones and have their water pumped for free. It's not just about work or cooking, however, it's about education too, a relief for students having to study at home by candlelight, and for people who can now charge phones, laptops and run other electrical devices. So with power comes information, and information means the people become more politically empowered; better education, better communication, better human rights and more democracy.

- In New Zealand, a Māori company called Kono is investing in regenerative farming as part of its sustainable production goals, including cutting down on agricultural waste. While they are referred to as a 'family business', this actually means the Māori community who make up the supply chain, including the farmers and fishermen, who have stakes in the company. This collective owns vineyards, orchards, oyster and mussel farms, fishing boats and a craft brewery, and they combine these resources; for example, they use the hops waste left over from brewing beer and crushed mussels shells for compost in their vineyards.

- Earth isn't just used for farming. Most modern construction involves cement, which produces masses of carbon dioxide, but in some countries, buildings have been constructed without cement that have lasted centuries. The Hakka people of the Fujian Province in China use a process called 'rammed earth', soil compressed a layer at a time and built up in slabs. Because it's not as stable as concrete, the walls have to be much thicker, which also means they offer much more insulation. There are versions of this type of traditional building technique in other places around the world, and it is now being used in an array of forms in modern architecture. As the global population grows, we have to find ways of constructing buildings that do less damage to our environment. In the Cloughjordan ecovillage, in Tipperary, all sorts of materials have been combined to build the different houses, including rammed earth, cob (a clay and straw mix), lime and hemp plaster and sheep's wool insulation.

Less is more

It goes without saying that the more *humans* there are on the planet, the more we do ... well, *everything*. I was born in 1973, and *within my lifetime*, the population of the world has *doubled*, from 3.7 billion to 7.5 billion. It is expected to reach 9.7 billion in 2050 and could peak at nearly 11 billion around 2100.

That's a lot of people, though the threat of it is often overstated. Over-population is blamed for a shortage of resources, but it's not that we don't have enough land or food, it's that often it's not distributed fairly.

If you are so poor and so lacking in support from society that your only form of security is the size of your family, and if your children are at risk of dying before they reach adulthood, you need a big family. The answer to our rising population then, along with sharing resources more fairly, is to ensure we raise the living standards of those in poverty. Though there are a lot of things that affect population growth, one of the most fundamental factors is women's rights. Rather than trying to use clumsy or cruel laws to limit the number of children people have, it's been shown that when women have better education and opportunities, they can plan and choose to have children later, and have fewer of them. This has been happening in Ireland, which was once one of the poorest countries in Europe. Gone are the days of regularly cramming eight or nine people into a five-seater car (with not a seatbelt between them). The average size of our families has dropped within a few generations, with the rise of wealth and women's rights. Some research suggests this is also taking effect worldwide, and that despite the current increase, the Earth's population will soon start flattening out, and by the year 2100 may even start to decline.

Money talks

Individuals can only do so much. We have to take on the giants who are the main cause of all this. And we are. A report in 2017 claimed that just *one hundred* active fossil-fuel producers, including ExxonMobil, Shell, BHP Billiton and Gazprom, were linked to 71 per cent of industrial greenhouse gas emissions since 1988. These companies have enormous power and influence, and if we are to have any chance of turning the climate crisis around, they have to be involved. While there may be plenty of well-intentioned people working for these multinationals, the companies themselves exist to make money, and losing money is the only sure way to get them to behave well.

Thankfully, this has already started happening. Here's one example. In 2019, there was a meeting of major investors in Cape Town in South Africa. These investors were the companies who put money into businesses so that they (the investors as well as the companies) can make a profit. Even big businesses like oil companies need other people's money, and they can collapse if the investors start taking their money out and putting it somewhere else. Between them, the firms that met in Cape Town controlled $11 trillion. Together, they decided that they would no longer invest in fossil fuels.

They're by no means the first – this has been going on for some time. In fact, Ireland passed a law in 2018 to ensure that public money could no longer be invested in fossil fuels. We were the first country in the world to do that. We also made headlines in 2017 when we banned fracking, that particularly destructive form of extracting oil or gas from the ground. And this action didn't come out of nowhere. People from community groups all over Ireland contacted their politicians, through email, phone and meetings, and asked them to put forward these two pieces of legislation. Big business was actually pushed out. These actions are both examples of politicians acting in response to their constituents. And as a result, big businesses started paying attention to the law, the science, the moral arguments ... and the sums of money they were going to lose long-term if they kept backing a doomed industry.

Talking about this issue matters. Demanding action from our governments matters. Now, all around the world, financial investors in control of enormous amounts of money are deciding that there's no future in burning stuff.

Engineering the Earth

No one solution will do. Cutting back on our greenhouse gas emissions is not going to be enough to slow our changing climate in time to make a real difference. To make that happen, we also need to do something about the atmosphere we have now. We've looked here at some of the more natural ways of doing this already, and there are other approaches that could work, referred to as 'geo-engineering'. Most of them are experimental and have never been tried on a large scale, which means there are risks involved, and we don't know how effective they would be. And the bigger they are, the riskier they get – we'd be interfering with an insanely complex system that we've already messed up. However, anything that could slow down what's happening could make a major difference to our future. Here are just a few of them.

Aerosol injection

When a volcano erupts, it blasts massive amounts of sulphur dioxide into the atmosphere. This can have the effect of scattering light and causing global cooling. There have been proposals to shoot this same stuff – sulphur dioxide – into the sky with cannons, or drop it from balloons or aircraft. It would take a whole *lot* of sulphur dioxide to have any effect, though, and unfortunately, it's also a pollutant itself and can result in acid rain. Partly for this reason, it would only be a temporary fix. We couldn't keep doing it. But it's possible that it might slow things down and give us time to make other solutions work.

On the Tibetan plateau, China's Aerospace Science and Technology Corporation is already using machines that shoot silver iodide into the atmosphere, which helps form clouds, in order to increase rainfall in the region. China's neighbours, particularly India, have mixed feelings about the eastern super-power messing with their sky, but it is proof of just how potent the technology can be.

Marine cloud brightening

Here's another idea people have had about clouds. Sea water could be blasted, from towers, up to a height of three hundred metres or more, which could help to create clouds, which in turn would help to reflect sunlight. This could be done on land, but it would be more practical to do it over the sea, using ships. This has the advantage of being able to target particular areas and it does not use any pollutants. But it would stop sunlight reaching the Earth and that could have consequences for plant growth, food production and our own mood, living under constant cloudy skies.

Ocean mirror

Because it's white, ocean foam reflects ten times more light than dark sea water, much in the way that sea ice does. One plan is for a fleet of ships that churn up the water, creating a temporary very large bright surface. This would take a lot of energy, and, obviously, you would have to be careful what you chose to power the ships.

Artificial trees

Rather than reducing the amount of sunlight reaching the planet's surface, another option might be to take carbon dioxide out of the air. Some power plants already have scrubbers on their chimneys to remove greenhouse gases from the smoke, but it's possible to go further and suck carbon dioxide from the atmosphere itself. One approach is to use artificial trees that absorb carbon dioxide at a thousand times the rate of real trees, using a process powered by solar cells. The carbon can then be stored in empty oil wells or other cavities below ground. It could even be used to make things. Carbon fibres are used in electronics, batteries, and aircraft and car components.

Ocean fertilisation

Remember the phytoplankton, that microscopic plant life that lives in the oceans and supplies us with so much of our oxygen? There's a theory that if you seed sections of the oceans with iron or nitrates, you can encourage phytoplankton to bloom and it will absorb more carbon dioxide, most of which will sink to the ocean floor when the phytoplankton dies, trapping it there. This theory has not been tested on a large scale, so we don't know how effective it would be, and because it means messing with ecosystems, there are concerns that it might do more harm than good.

None of these ideas is a replacement for the basics; we have to stop pumping greenhouse gases into the atmosphere. But there is no single solution to climate change; it's going to take a whole range of actions, some of which will be new, unconventional and might even seem slightly bonkers. We need to keep our minds open to the possibilities. And we need hope.

Friends of the Earth Ireland

Ireland's fracking ban happened thanks to communities taking a stand against dirty energy to protect the environment and climate. They got organised, lobbied the government and protested. Today these communities continue to fight in solidarity with other communities that face fracking, such as in Northern Ireland and the USA.

CHAPTER 22
REASONS FOR HOPE

There are as many ways to help tackle climate change as there are problems that result from it. When you look at all the stupid things that humans do – the petty arguments, the selfishness and corruption, the devastating wars – it can seem like we'll never be able to work together in the face of this crisis, but we must remind ourselves that we have also achieved extraordinary things, so in this last chapter I want to give some examples of times when humans got together and agreed to work together, for the sake of the whole planet.

It's something we're doing more often as time goes on, and we're getting better at it – and *faster* too.

In 1903, the Wright brothers made the first powered flight, covering less than a hundred metres. In 1919, Alcock and Brown made the first non-stop flight across the Atlantic in a plane made mostly from wood, wire and varnished linen. This was an awesome achievement for the time (I wrote a novel about it), and they couldn't have done it without the aircraft manufacturer and a range of skilled professionals.

In just *fifty years*, we went from barely making it across the Atlantic Ocean in a wooden biplane to landing a human being on the moon. That's all it took, fifty years. And those three guys in the space capsule didn't do it on their own, of course. They were supported by *400,000 people*, numerous private companies, and science developed over thousands of years from around the world. Which brings us to the moon.

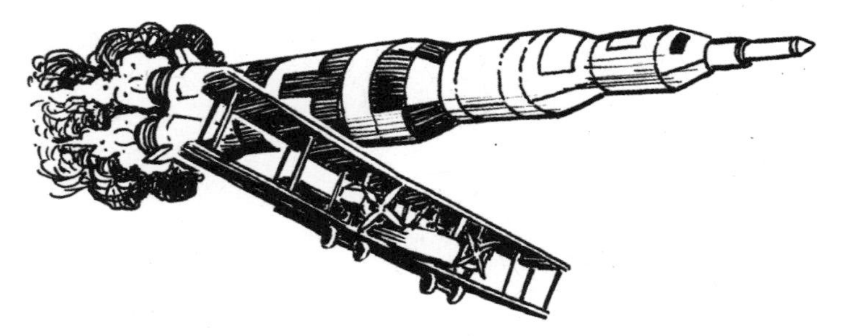

The moon

I mentioned the Antarctic Treaty earlier. If Antarctica was likely to start arguments over who owned it, you can imagine there'd be similar issues over who got to claim the *moon*. Despite the fact that the United States was the first country to stick a flag on it, it remains neutral territory. The Moon Agreement was adopted by the United Nations in 1979, although it didn't come into force until 1984. It also applies to other celestial bodies. Simply put, the moon belongs to all of humankind. You are one of the owners of the moon. Congratulations.

The eradication of smallpox

Smallpox was a horrible disease that killed about 30 per cent of the people it infected. Millions died of it every year, and many more were left permanently disfigured.

There was no treatment once you were infected. You'll notice that I'm referring to this disease in the past tense. For the first half of the 20th century, a mass vaccination campaign helped reduce the number of deaths, but smallpox remained a serious problem in South America, Africa and Asia. Then in 1966, the World Health Organisation started a programme to wipe the disease out, permanently. Instead of just waiting for people to come and get vaccinated, *they searched for every single case of the disease*. It was a colossal operation, involving everyone from national governments to local health workers. In 1980, the WHO declared that the disease had been eradicated. There wasn't a single case of smallpox left on Earth. The WHO has now set its sights on polio, and again, it is a huge international effort.

The Millennium Bug

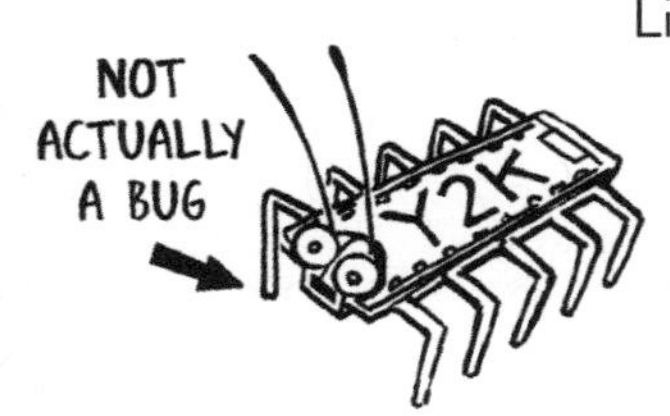

Little changes can have big effects. Back in the early days of computers, storage space was precious – and expensive – and for companies that dealt with a lot of data, even cutting down on numbers or letters in data could save on memory and money. Dates were repeated a lot in computing so, to save space, they were recorded with only two figures for the year, like 25/12/75. This made sense at the time, and decades went by without any problems. By the 1990s, computers were used to run everything from international banking to air traffic control systems, from hospitals to power plants. But as the year 2000 approached, programmers were realising they had a problem.

With only two digits denoting the year, how would all these computers be able to tell the difference between the first of January 2000 (1/1/00) and the first of January 1900 (1/1/00)? What would happen to all these automated systems the world relied on when their calendars told them the date had been reset to the start of the century? Nobody knew. Nothing like this had ever happened before. It could cause a worldwide crisis. Not only were so many essential services running on computers, but each service was tied into *other* systems that had to keep working properly. But as the calendars clicked to the year 2000, it *didn't* cause a worldwide crisis. Because of this, some people think that the Millennium Bug, or Y2K Bug, was a false alarm, or even a complete hoax. What they don't realise is that programmers all around the world anticipated the problems and worked to prevent the crisis from happening by making changes to both software and hardware. The total cost of preventing the Millennium Bug is thought to have been somewhere between $300 billion and about $460 billion.

The hole in the ozone layer

OK, it wasn't actually a hole. The ozone layer is a part of the atmosphere that acts as a natural sunscreen, protecting life on Earth by absorbing ultraviolet light, which damages DNA in plants and animals (including ...

y'know, humans) and leads to sunburn and skin cancer. Back in the 1980s, scientists measuring concentrations of the gas in the atmosphere realised that it was thinning out over the South Pole. It was all the fault of chlorofluorocarbons – though everyone called them CFCs, because, I mean 'chlorofluorocarbons'? Really?

These chemicals that we had been using in refrigerators and aerosol sprays since the 1930s reacted with the ozone molecules, causing them to break apart. In a remarkable feat of science communication, these experts actually convinced everyone that skin cancer was bad, that we messed with the ozone layer at our peril, and we had to step up and deal with this thing. This led to the Montreal Protocol in 1987, a treaty that phased out the production of chemicals that thinned out the ozone. Recent evidence suggests that the hole has actually started closing. Imagine that; people actually listening to scientists.

The world wide web

It's hard to imagine a world without the web now, but it had to start somewhere, and it started at CERN, the European Council for Nuclear Research, a laboratory complex that sits astride the Franco-Swiss border near Geneva. It was the brainchild of a man named Tim Berners-Lee in 1989. There were a lot of different computer systems being used for different projects in CERN, and they couldn't communicate with each other. People were having trouble sharing information. It was all getting very awkward. The internet already existed – computers connected by phones – but they didn't all speak the same language.

It was Berners-Lee who created HyperText Markup Language (HTML), which allowed all the different computers to talk to each other, the Uniform Resource Identifier (URI, now called the URL), which is a way for each computer to have an address on the web, and the Hypertext Transfer Protocol (HTTP), a way to link to resources, so you could reach out and find things. Crucially, he got CERN to agree that this language would be released *free to the whole world*, so that everyone could use it, and it just spread from there. Tim Berners-Lee doesn't get any money for having invented the worldwide web; he just did it as part of his job.

The nuclear arms treaty

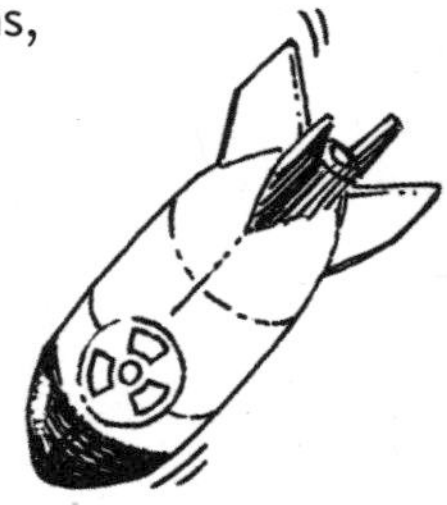

Back during the Cold War, in the 1950s and 1960s, everyone wanted nuclear weapons, and the United States and Russia were so intent on having the most that, between them, they built enough to wipe out all life on Earth many times over. Everyone agreed that it was all getting a bit daft and, sooner or later, something was going to go wrong somewhere, and *we were all going to die*. It was time for all the countries concerned to *calm down*. And so the Treaty on the Non-Proliferation of Nuclear Weapons was brought into force in 1970. A total of 191 states have joined, committing to prevent the spread of nuclear weapons and work towards getting rid of them altogether. It should be pointed out that there are still plenty of nuclear weapons around, but it did help cool things down, and, well ... it's a work in progress.

The COVID-19 pandemic

When it became clear in late 2019 that, with COVID-19, the world was facing a virus that was infectious and dangerous enough to create a pandemic, there was a global response. Scientists had mapped the virus's genetic structure by January 2020, and shared it across the world, allowing tests to be developed and vaccine research to begin, even as countries were closing their borders and restricting travel and social contact to protect their citizens.

The World Health Organisation immediately began gathering the latest scientific findings and knowledge on the disease and compiling it in a database for use by medical professionals everywhere. It brought together scientists, researchers and national public health experts from around the globe to figure out the best ways to prevent and fight the virus. The WHO also committed to ensuring that, as medicines and vaccines were developed, the science would be shared fairly with all countries and people. It was universally acknowledged in the medical community that the only way to eradicate this virus *anywhere* was to stop it *everywhere*. Less than a year after the threat of the virus had been recognised, the first successful vaccines were announced – a process that can often take as long as ten years. Because of the international co-operation, they had been developed at unprecedented speed.

The Paris Agreement

In December 2015, the Paris Agreement became the first-ever universal, legally binding global climate change agreement. To come into force, it needed at least fifty-five countries representing at least 55 per cent of global emissions to sign up. To date, 197 countries have confirmed their agreement. It's a long, complicated legal document that covers every element of the climate crisis, but the bones of it is that these countries have set themselves the target of keeping the rise in global temperatures this century to well below 2° Celsius above pre-industrial levels and to do their best to limit the temperature increase even further to 1.5° Celsius. It's a whole-world approach to a whole-world problem, and includes helping countries adapt to the impacts of climate change, with industrialised countries committing support to their less wealthy neighbours.

These kinds of complex legal documents can seem drearily political, theoretical ... boring. But the actions of governments are steered by laws. This agreement represents the power of most of the nations of the world, sharing knowledge, expertise, money and resources in an act of global co-operation. It's a statement that says: 'Your problems are my problems. We have to fix this together.'

You are part of this story

This is such a big story, and we've only been able to touch on some of the many important subjects here, but I'd urge you to do your own exploring. Don't rely on what I've said. Check the facts out for yourself, and find the things that fascinate you. With this book, I wanted to show how human civilisation is woven into the environment, that they are not separate entities, and that each of us is part of the awesome beauty and complexity of that system. Billions of years of evolution have led to the person you are now. You are a marvel of nature, linked to your environment through air and water and food and other conditions that are so finely balanced, so rare in the universe, that it's astonishing that life can exist at all. Each life is something remarkable, and has an effect on the world around it. And *what a world this is*! Look at the wonders it offers. You could live a hundred lifetimes and never be able to take it all in, but I hope, like me, that you gaze around from time to time in utter amazement at what we have. There is magic in the tiniest details.

We are part of a thin skin of life around a ball of rock, floating in space. Our control over our environment made us powerful; it allowed us to become the most advanced creatures on the planet. But for too long, we've been getting ahead of ourselves, charging forward so fast that we can't see where we're going. Now it's time to recognise what it took to get us here, the price the Earth has paid, and to start putting more thought into our future.

It's time to take better care of our world. It's fragile, complicated and majestic ...

... And it's the only one we have.

Friends of the Earth Ireland

The biggest changes in history have been made by regular people like you and me – by coming together, standing up for what is right and demanding change. Civil rights, marriage equality and the suffragettes are just some of the movements that came before us, and today the climate movement is vibrant and strong. People Power can change the world.

ACKNOWLEDGEMENTS

Many people helped shape the content and narrative of this book by reviewing the text at various stages and sharing their expertise. I'd like to thank Matthew, Siobhán and Kate at Little Island, and Emma Dunne for the proofreading. I'm very grateful to the Friends of the Earth team, in particular Kate Ruddock, Claudia Tormey, Oisín Coghlan, Áine O'Gorman and Deirdre Duff for their input, and to all the people who were so kind in donating to the crowdfunder for the book's production.

Friends of the Earth would also like to express huge thanks to their supporters for donating to help fund the making of this book. Thanks to their former chair and climate scientist Cara Augustenborg, to their former board secretary and author Anna Heussaff and to Professor John Sharry, founder of Parents Plus and chair of Feasta.

Special thanks to the young activists across Ireland for their involvement and inspiration, especially Aiyana Hedler and David Mallin.

NOTE

We only scratch the surface of the topics in this book. For further reading and for many of Oisín's research links, check out the page for this book on the Little Island website: www.littleisland.ie.

Whatever you're into, there's more to explore.

DONORS

The following people helped make this book possible with donations of €100 or more:

Andrew Owen-Griffiths
Anita Wheeler
Anne Harnett
Barbara Bracken
Bernadette Power
Brian and Eileen Coates
Caragh Behan
Celia Keenan
Dave Linehan
David Hutchinson Edgar
Derek Cawley
Donal Daly
Edmond Grace
Emer Lawlor
Emma Lane-Spollen
Eoin Colfer
Frank Murphy
Frank McDonald
Graham Lightfoot
Helen McCauley
Jeannette Golden
John Heffernan
Kate Wheeler
Ken Jordan
Kunak McGann
Jochen Gerz and
Laurence Vanpoulle
Mary Jo Quigley
Oisin Coghlan
Pat O'Gorman
Richard Moles
Ronan Moore
Ronan Desmond
Rosalind Duke
Ruth Blackith
Sean Dunne
Tom Roche,
Founder of Just Forests
Eileen McGann

ABOUT THE AUTHOR

Oisín McGann is a best-selling and award-winning writer and illustrator. He has produced dozens of books and short stories for all ages of reader, including twelve novels. In 2014 and 2015, he was the Irish writer-in-residence for Weather Stations, an EU-funded project where writers from five different countries were tasked with finding ways to use storytelling to raise awareness of climate change. He has carried on this work through school residencies in primary and secondary schools, funded by Poetry Ireland Writers in Schools and Irish Aid's WorldWise Global Schools. He is married with three children, two dogs and a cat, and lives somewhere in the Irish countryside, where he won't be heard shouting at his computer.

www.oisinmcgann.com

ABOUT FRIENDS OF THE EARTH IRELAND

Friends of the Earth Ireland campaign and build movement power to bring about the system change needed for a just world where people and nature thrive. We support people to come together to transform our world until social justice is the foundation of resilient and regenerative societies that flourish within the ecological limits of our one planet.

www.friendsoftheearth.ie

ABOUT LITTLE ISLAND BOOKS

Little Island publishes good books for young minds, from toddlers to teens. Based in Dublin, Ireland, Little Island specialises in Irish writers and also publishes some works in translation. It is Ireland's only English-language publisher that specialises in books for young readers, and receives funding from the Arts Council of Ireland.

www.littleisland.ie